Signals and Communication Technology

More information about this series at http://www.springer.com/series/4748

Kandarpa Kumar Sarma
Manash Pratim Sarma · Mousmita Sarma
Editors

Recent Trends in Intelligent and Emerging Systems

Springer

Editors
Kandarpa Kumar Sarma
Department of Electronics and Communication Technology
Gauhati University
Guwahati, Assam
India

Mousmita Sarma
Department of Electronics and Communication Engineering
Gauhati University
Guwahati, Assam
India

Manash Pratim Sarma
Department of Electronics and Communication Engineering
Gauhati University
Guwahati, Assam
India

ISSN 1860-4862 ISSN 1860-4870 (electronic)
Signals and Communication Technology
ISBN 978-81-322-2406-8 ISBN 978-81-322-2407-5 (eBook)
DOI 10.1007/978-81-322-2407-5

Library of Congress Control Number: 2015937946

Springer New Delhi Heidelberg New York Dordrecht London

Printed on acid-free paper

Springer (India) Pvt. Ltd. is part of Springer Science+Business Media
(www.springer.com)

The volume is dedicated to students, research scholars, faculty members of the Electronics fraternity of Gauhati University and everyone associated with intelligent system design.

Preface

The present volume *Recent Trends in Intelligent and Emerging System Design* to be published in the Signals and Communication Technology Series of Springer is a compilation of the recent trends in research and development in the related areas.

The book attempts to cover trends in intelligent and emerging system design using a range of tools, including soft-computation. The book intends to include discussion and experimental work in the areas of communication, computation, vision sciences, device design, fabrication, upcoming materials and related process design, etc. Its objective is to provide a glimpse to the reader of the recent trends in research in the areas of intelligent and emerging system design and related areas. The book shall be a valuable compilation of recent research works in these areas. The contents are likely to encourage young science and engineering pass-outs to consider research a viable career option.

The audience of this book is students and researchers who deal with intelligent and emerging system design through mathematical and computational modeling and experimental designs. Specifically, audiences that are broadly involved in the domains of electronics and communication, electrical engineering, mathematics, computer science, other applied informatics domains, and related areas.

The work included in the book broadly covers all areas of Electronics and Communication Engineering and Technology, Soft-computational applications, Human Computer Interaction (HCI) Designs, and Social and Economic Dynamics. Certain works included in the work reflect the current trends in design and development. The works included in the volume has been grouped into Intelligent Applications in Communication, Selected Issues in Biomedical and Social Science, HCI and Bio-inspired System Design, Soft Computing and Hybrid System-based Speech Processing Applications, and Review Chapters on Selected Areas. These groupings constitute five different segments. In Part I, five works covering trends in communication system design have been included. Two works constitute Part II, where contributions related to HCI and bio-inspired designs have been included. In the next part, there are works related to biomedical and social science domains. Here, three works are included. In Part IV, works related to speech processing, especially that of recognition and synthesis are included. The final part is based on three review papers.

The works of Basab Bijoy Purkayastha et al., Hemashree Bordoloi et al., etc., reflect innovative applications of already known approaches of system design. Works by Jumi Kalita et al. and Deepak Goswami et al. represent applications of soft computing tools for social issues and related dynamics. Some of the familiar human–computer interaction applications are highlighted using speech and gesture recognition-based designs by Pallabi Talukdar et al. and Dharani Mazumdar et al. All papers have been passed through multiple rounds of review, modification, and correction in order to ensure quality in the finally compiled form. The editors expect the compilation will be interesting to read.

The editors are thankful to the contributors, reviewers, Springer, and the series editor for making the compilation possible.

Guwahati, Assam, India
February 2015

Kandarpa Kumar Sarma
Manash Pratim Sarma
Mousmita Sarma

Acknowledgments

The editors acknowledge the contributions by the authors, reviewers, students, researchers, and faculty members who have contributed in their respective ways in making the compilation of the volume possible.

Contents

Editors and Contributors

About the Editors

Dr. Kandarpa Kumar Sarma currently Associate Professor in Department of Electronics and Communication Technology, Gauhati University, Guwahati, Assam, India, has over seventeen years of professional experience. He has covered all areas of UG/PG level Electronics courses including Soft computing, Mobile communication, Pattern Recognition, Digital signal, and image processing. He obtained the M.Tech. degree in Signal Processing from Indian Institute of Technology Guwahati, Guwahati, Assam, India in 2005. Subsequently he completed Post Doc. Research from Technical University Sofia, Bulgaria in 2015. He has authored/ co-authored/ edited eleven books, several book chapters, and several peer-reviewed research papers in international conference proceedings and journals. His areas of interest include Soft-Computation and its Applications, Knowledge aided System Design, MIMO, Mobile Communication, Antenna Design, Speech Processing, Document Image Analysis and Signal Processing Applications in High Energy Physics, Neuro-computing and Computational Models for Social-Science Applications. He is senior member IEEE (USA), Fellow IETE (India), Member International Neural Network Society (INNS, USA), Life Member ISTE (India), and Life Member CSI (India). He serves as an Editor-in-Chief of International Journal of Intelligent System Design and Computing (IJISDC, UK), guest editor of several international journals, reviewer of over thirty international journals, and reviewer/TPC member of over hundred international conferences.

Manash Pratim Sarma is presently working as Assistant Professor in the Department of Electronics and Communication Engineering, Gauhati University, Assam, India. He earned his M.Sc. in Electronics Science from Gauhati University, Assam, India in 2008. He also received the M.Tech. in Electronics Design and Technology from Tezpur University, Tezpur, Assam, India in 2010. He has authored several peer-reviewed publications in International journals, International and National Conference proceedings. He has also served as reviewer in many international conferences and a few journals. His areas of interests include Wireless communication and VLSI design for high data rate communications.

Ms. Mousmita Sarma is currently working as Research Consultant at M/s. SpeecHWareNet Pvt. (I) Ltd., Technology Incubation Center, IIT Guwahati, Assam, India. She completed her M.Sc. in Electronics and Communication Technology from Gauhati University, India in 2010. She also completed the M.Tech. from the same institution in 2012 with specialization in Speech Processing and Recognition. She has co-authored two books and published several peer-reviewed research papers in international conference proceedings and journals. She serves as reviewer to several journals and IEEE international and national conferences. Her areas of interest include Speech Recognition, Soft-Computation, and HCI Applications.

Contributors

Utpal Bhattacharjee Department of Computer Science and Engineering, Rajiv Gandhi University, Doimukh, Arunachal Pradesh, India

Manasjyoti Bhuyan Department of Electronics and Communication Technology, Gauhati University, Guwahati, Assam, India

Hemashree Bordoloi Department of Electronics and Communication Engineering, Assam Don Bosco University, Guwahati, Assam, India

Atlanta Choudhury Department of Electronics and Telecommunication Engineering, GIMT, Guwahati, Assam, India

Banti Das Department of Electronics and Communication Technology, Gauhati University, Guwahati, Assam, India

Krishna Dutta NIT Nagaland, Dimapur, Nagaland, India

Deepak Goswami Department of Economics, Lalit Chandra Bharali College, Guwahati, Assam, India

Nisha Goswami Department of Electronics and Communication Technology, Gauhati University, Guwahati, Assam, India

Pradyut Kumar Goswami Assam Science and Technology University, Guwahati, Assam, India

Padma Lochan Hazarika Department of Commerce, Gauhati University, Guwahati, Assam, India

Jumi Kalita Department of Statistics, Lalit Chandra Bharali College, Guwahati, India

Dharani Mazumdar Department of Electronics and Communication Engineering, Gauhati University, Guwahati, Assam, India

Debashis Dev Misra Royal School of Engineering and Technology, Guwahati, Assam, India

Madhurjya Kumar Nayak Department of Electronics and Communication Engineering, Gauhati University, Guwahati, Assam, India

Basab Bijoy Purkayastha Department of Physics, IIT Guwahati, Guwahati, Assam, India

Samar Jyoti Saikia Department of Electronics and Communication Engineering, Assam Don Bosco University, Guwahati, Assam, India

Dipjyoti Sarma Department of Electronics and Communications Engineering, Don Bosco College of Engineering and Technology, Assam Don Bosco University, Guwahati, India

Kandarpa Kumar Sarma Department of Electronics and Communication Technology, Gauhati University, Guwahati, Assam, India

Manash Pratim Sarma Department of Electronics and Communication Engineering, Gauhati University, Guwahati, Assam, India

Mousmita Sarma Department of Electronics and Communication Engineering, Gauhati University, Guwahati, Assam, India

Pranita Sarmah Department of Statistics, Gauhati University, Guwahati, India

Anjan Kumar Talukdar Department of Electronics and Communication Engineering, Gauhati University, Guwahati, Assam, India

Pallabi Talukdar Department of Electronics and Communication Engineering, Gauhati University, Guwahati, Assam, India

Tapashi Thakuria Department of Electronics and Communication Engineering, Gauhati University, Guwahati, Assam, India

Acronyms

AANN — Autoassociative Neural Network: Autoassociative Neural Networks are feedforward nets trained to produce an approximation of the identity mapping between network inputs and outputs using backpropagation or similar learning procedures.

AD — Alzheimer's Disease: Alzheimer's Disease is the most common form of dementia, a general term for memory loss and other intellectual abilities serious enough to interfere with daily life.

ADC — Analog to Digital Converter: Analog to Digital Converter is a device that converts a continuous physical quantity to a digital number

ANFCE — Adaptive Neural Fuzzy Channel Equalizer: Adaptive Neural Fuzzy Channel Equalizer is form of channel equalizer designed using ANFIS.

ANFIS — Adaptive Neuro Fuzzy Inference System: Adaptive Neuro Fuzzy Inference System is a kind of fuzzy and artificial neural network that is based on Takagi-Sugeno fuzzy inference system.

ANN — Artificial Neural Network: Artificial Neural Networks are non-parametric models inspired by animal central nervous systems. These are particularly the brain that are capable of machine learning and pattern recognition and usually presented as systems of interconnected "neurons" that can compute values from inputs by feeding information through the network.

AP — Amyloid Proteins: Amyloid Proteins are insoluble fibrous protein aggregates sharing specific structural traits.

ARPA — Advanced Research Projects Agency: Advanced Research Projects Agency is an agency of the United States Department of Defense responsible for the development of new technologies for use by the military.

ASK — Amplitude-Shift Keying: Amplitude-Shift Keying is a form of modulation where a digital data stream modulates the amplitude of a continuous wave.

ASR Automatic Speech Recognition: Automatic Speech Recognition is a technology by which a computer or a machine is made to recognize the speech of a human being.

ATM Asynchronous Transfer Mode: Asynchronous Transfer Mode is defined for carriage of a complete range of user traffic, including voice, data, and video signals developed to meet the needs of the Broadband Integrated Services Digital Network, which was designed for a network that must handle both traditional high-throughput data traffic (and real-time, low-latency content).

AWGN Additive White Gaussian Noise: Additive White Gaussian Noise is a channel impairment to communication which is additive and shows a constant spectral density and a Gaussian distribution of amplitude.

BER Bit Error Rate: Bit Error Rate is the number of received bits of a data stream over a communication channel that have been altered due to noise, interference, distortion or bit synchronization errors.

BIST Built-in Self Test: Built-in Self Test or built-in test is a mechanism that permits a machine to test itself mends for high reliability and lower cycle times.

BMI Body Mass Index: Body Mass Index or Quetelet index, is a measure for human body shape based on an individual's mass and height.

BPA Back Propagation Algorithm: Back Propagation Algorithm is a common supervised method of training ANN where from a desired output, the network learns from many inputs.

BPSK Binary Phase Shift Keying: Binary Phase Shift Keying is a type of phase modulation using two distinct carrier phases to signal ones and zeros which is the simplest form of PSK.

BPTT Back Propagation Through Time: Back Propagation Through Time is a gradient-based technique for training certain types of Recurrent Neural Network (RNN)s.

CBLRNN Complex Bilinear Recurrent Neural Network: Complex Bilinear Recurrent Neural Network is form of Recurrent Neural Network (RNN).

CCI Co Channel Interference: Co Channel Interference is crosstalk from two different radio transmitters using the same frequency.

CCS Code Composer Studio: Code Composer Studio is an integrated development environment (IDE) for Texas Instruments (TI) embedded processor families.

CDMA Code Division Multiple Access: Code Division Multiple Access is a code based channel access method used by various radio communication technologies where an user can access the whole bandwidth can be used by the user all the time.

CMN	Cepstral Mean Normalization: In ASR technology Cepstral Mean Normalization is a method used to minimise the effect of channel characteristics differences which may vary from one session to the next. It involves subtracting the cepstral mean, calculated across the utterance, from each frame.
CMOS	Complementary Metal Oxide Semiconductor: Complementary Metal Oxide Semiconductor refers to the fact that the typical digital design style with CMOS uses complementary and symmetrical pairs of p-type and n-type metal oxide semiconductor field effect transistors (MOSFETs) for logic functions.
CNR	Carrier-to-Noise Ratio: Carrier-to-Noise Ratio is the ratio of the received modulated carrier signal power to the received noise power in a specific bandwidth.
CSI	Channel State Information: Channel State Information refers to known channel properties of a communication link.
CTDFRNN	Complex Time Delay Fully Recurrent Neural Network: Complex Time Delay Fully Recurrent Neural Network is a special form of Recurrent Neural Network (RNN).
DAC	Digital to Analog Converter: Digital to Analog Converter is a device that converts a digital (usually binary) code to an analog signal.
DEKF	Decoupled Extended Kalman Filter: Decoupled Extended Kalman Filter is a technique used to train Recurrent Neural Network (RNN) with separated blocks of Kalman filter.
DEPSO	Differential Evolution PSO: Differential Evolution PSO is an optimization technique.
DIBL	Drain Induced Barrier Lowering: Drain Induced Barrier Lowering is a short-channel effect in MOSFETs referring originally to a reduction of threshold voltage of the transistor at higher drain voltages.
DLL	Delay Locked Loop: Delay Locked Loop is a digital circuit similar to a phase-locked loop (PLL), with the main difference being the absence of an internal voltage-controlled oscillator, replaced by a delay line.
DPLL	Digital Phase Locked Loop: A Digital phase-locked loop is a digital control system that generates an output signal whose phase is related to the phase of an input signal.
DPSK	Differential Phase Shift Keying: Differential Phase Shift Keying or dual-polarization QPSK involves the polarization multiplexing of two different QPSK signals for improving the spectral efficiency.
DSK	Digital Signal Processing Kit: Digital Signal Processing Kit is an electronic board with Digital Signal Processor used for experiments, evaluation and development.
DSP	Digital Signal Processing: Digital Signal Processing is the mathematical manipulation of an information signal to modify or improve it in some way.

DS-UWB	Direct Sequence-Ultra Wideband: Direct Sequence-Ultra Wideband is often referred to as an impulse, baseband or zero carrier technology which operates by sending low power Gaussian shaped pulses which are coherently received at the receiver.
DVS	Dynamic Voltage Scaling: Dynamic Voltage Scaling is a power management technique in computer architecture, where the voltage used in a component is increased or decreased, depending upon circumstances.
EA	Evolutionary Algorithm is a subset of evolutionary computation, a generic population-based metaheuristic optimization algorithm.
ECC	Elliptic Curve Cryptosystem is an approach to public-key cryptography based on the algebraic structure of elliptic curves over finite fields.
EEA	Extended Euclidean Algorithm is an extension to the Euclidean algorithm. Besides finding the greatest common divisor of two integers, as the Euclidean algorithm does, it also finds other two integers (one of which is typically negative) that satisfy Bézout's identity.
EGC	Equal Gain Combining: Equal Gain Combining is a form of diversity.
EM	Expectation-Maximization: Expectation-Maximization is an iterative method for finding maximum likelihood or maximum a posteriori (MAP) estimates of parameters in statistical models, where the model depends on unobserved latent variables.
FCM	Fuzzy C-means: Fuzzy C-means is a form of clustering technique based on fuzzy logic.
FCRNNs	Fully Connected RNNs is a network of RNNs, each with a directed connection to every other unit.
FD	Factorial Design: Factorial Design is the most common way to study the effect of two or more independent variables, although we will focus on designs that have only two independent variables for simplicity.
FEOL	Front-End of Line is the first portion of IC fabrication where the individual devices (transistors, capacitors, resistors, etc.) are patterned in the semiconductor.
FIS	Fuzzy Inference System: Fuzzy Inference System is a system that uses fuzzy set theory to map inputs (features in the case of fuzzy classification) to outputs (classes in the case of fuzzy classification).
FL	Fuzzy Logic: Fuzzy Logic is a form of many-valued logic which deals with reasoning that is approximate rather than fixed and exact.
FN	Fuzzy Neural: Fuzzy Neural refers to combinations of ANNs and fuzzy logic.

FNS Fuzzy Neural System: Fuzzy Neural System is a learning machine that finds the parameters of a fuzzy system (i.e., fuzzy sets, fuzzy rules) by exploiting approximation techniques from neural networks.

GDALMBP Gradient Descent with Adaptive Learning Rate and Momentum Back Propagation.

GDALRBP Gradient Descent with Adaptive Learning Rate Back Propagation.

GDBP Gradient Descent Back Propagation: Gradient Descent Back Propagation is form of ANN training method.

GDLMBP Gradient Descent with Levenberg-Marquardt Backpropagation.

GDMBP Gradient Descent with Momentum Back Propagation.

GEKF Global Extended Kalman Filter: Global Extended Kalman Filter is an extension of Kalman filter.

GIDL Gate Induced Drain Leakage: Gate Induced Drain Leakage is the effect for which current arises in the high electric field under the Gate-Drain overlap region of a MOSFET causing deep depletion.

GMM Gaussian Mixture Model: Gaussian Mixture Model is a probabilistic model for representing the presence of subpopulations within an overall population, without requiring that an observed data-set should identify the sub-population to which an individual observation belongs where mixture distribution is Gaussian.

GPPC General-purpose Personal Computer: General-purpose Personal Computers ia a computer built around a microprocessor for use by an individual having capabilities to perform wide range of tasks.

HCI Human Computer Interface: Human Computer Interface involves the study, planning, and design of the interaction between people (users) and computers.

HMM Hidden Markov Models: Hidden Markov Models is a statistical Markov model in which the system being modeled is assumed to be a Markov process with unobserved or hidden states. It can be considered to be the simplest dynamic Bayesian network.

IC Integrated Circuit: Integrated Circuit is a set of electronic circuits embedded on one small chip of semiconductor material, normally silicon.

IEEE Institute of Electronics and Electrical Engineering: Institute of Electronics and Electrical Engineering is a professional association headquartered in New York City that is dedicated to advancing technological innovation and excellence.

IS Incipient Sickness: Incipient Sickness is a condition of industrial health.

ISI Inter Symbol Interference: Inter Symbol Interference is a form of distortion of a signal in which one symbol interferes with subsequent symbols. It arises in high data rate wireless system.

ITA Itoh-Tsujii algorithm: Itoh-Tsujii algorithm is a generic algorithm used to invert elements in a finite field.

ITU	International Telecommunication Union: International Telecommunication Union originally founded as the International Telegraph Union is a specialized agency of the United Nations that is responsible for issues that concern information and communication technologies.
JSL	Japanese Sign Language: Japanese Sign Language is the dominant sign language in Japan.
KMC	K-means clustering: K-means clustering is a method of vector quantization originally from signal processing, that is popular for cluster analysis in data mining.
LCR	Level Crossing Rates: The LCR for a specific threshold is the expected rate at which the normalized envelop of a signal passes the threshold with a positive slope.
LMS	Least Mean Square algorithms: Least Mean Square algorithms are a class of adaptive filter used to mimic a desired filter by finding the filter coefficients that relate to producing the least mean squares of the error signal.
LOS	Line of Sight: Line of Sight refers the straight path from the transmitter to the receiver in wireless communication.
LPC	Linear Prediction Coding: Linear Prediction Coding is a tool used mostly in audio signal processing and speech processing for representing the spectral envelope of digital signal of speech in compressed form, using the information of a linear predictive model.
LPCC	Linear Prediction Cepstral Coefficients: Linear Prediction Cepstral Coefficients are the coefficients that can be found by converting the Linear Prediction coefficients into cepstral coefficients.
LS	Least Square: The method of least squares is a standard approach to the approximate solution of overdetermined systems.
LSDPFF	Least Square Digital Polynomial Fitting Filters: A digital filter which follows least square polynomial.
LSPF	Least Square Polynomial Fitting: Least Square Polynomial Fitting is a polynomial based fitting that optimizes in least square term.
LTI	Linear Time Invariant: Linear Time Invariant refers to the behavior of a system whose response varies linearly with time.
LVQ	Learning Vector Quantization: Learning Vector Quantization is a prototype-based supervised classification algorithm and is the supervised counterpart of vector quantization systems.
MAP	Maximum Aposteriori Probability: Maximum Aposteriori Probability is a mode of the posterior distribution in Bayesian statistics which can be used to obtain a point estimate of an unobserved quantity on the basis of empirical data.
MCS	Monte Carlo Simulation: Monte Carlo Simulations is a computerized mathematical technique that performs risk analysis by building models of possible results by substituting a range of values-a probability distribution in quantitative analysis and decision making.

MEMM Maximum Entropy Markov Model: Maximum Entropy Markov Models maximum-entropy Markov model or conditional Markov model, is a graphical model for sequence labelling that combines features of hidden Markov models (HMMs) and maximum entropy models.

MFCC Mel Frequency Cepstral Coefficient: Mel Frequency Cepstral Coefficients are coefficients that collectively make up an Mel Frequency Cepstrum which is a representation of the short-term power spectrum of a sound, based on a linear cosine transform of a log power spectrum on a nonlinear mel scale of frequency.

MIMO Multiple-Input Multiple-Output: Multiple-Input Multiple-Output is the use of multiple antennas at both the transmitter and receiver to improve communication performance.

MLP Multi Layer Perceptron: Multi Layer Perceptron is a feedforward artificial neural network model that maps sets of input data onto a set of appropriate outputs.

MLSD Maximum Likelihood Sequence Detector: Maximum Likelihood Sequence Detector is a device which uses a mathematical algorithm to extract useful data out of a noisy data stream.

MMSE Minimum Mean Square Error: Minimum Mean Square Error is an estimation method in statistics and signal processing which minimizes the mean square error (MSE) of the fitted values of a dependent variable, which is a common measure of estimator quality

MPEG Moving Pictures Experts Group: Moving Pictures Experts Group is a working group of ISO/IEC with the mission to develop standards for coded representation of digital audio and video and related data.

MRA Multiple Regression Analysis: Multiple Regression Analysis regression analysis is a statistical process for estimating the relationships among variables.

MRC Maximal Ratio Combining: Maximal Ratio Combining is a method of optimum diversity combining for independent AWGN channels which can restore a signal to its original shape.

MSB Most Significant Bit: Most Significant Bit is the bit position in a binary number having the greatest value.

MSE Mean Square Error: Mean Square Error is one of many ways to quantify the difference between values implied by an estimator and the true values of the quantity being estimated.

MSE Mean Square Equalizer: Mean Square Equalizer is an equalization method that works in the least square sense.

MSI Multi Stream Interference: Multi Stream Interference is a phenomenon where the different transmitted streams from each antenna in a multiple input case interfere at the receiver.

NCO Numerically Controlled Oscillator: Numerically Controlled Oscillator is a digital signal generator which creates a synchronous, discrete-time, discrete-valued representation of a waveform.

NF Neuro Fuzzy: Neuro Fuzzy refers to combinations of artificial neural networks and fuzzy logic.

NFS Neuro Fuzzy System: Neuro Fuzzy System is a learning machine that finds the parameters of a fuzzy system (i.e., fuzzy sets, fuzzy rules) by exploiting approximation techniques from neural networks.

NIPCCD National Institute of Public Cooperation and Child Development: National Institute of Public Cooperation and Child Development is an Indian government agency in New Delhi under the Ministry of Women and Children Development tasked with promotion of voluntary action research, training and documentation in the overall domain of women empowerment and child development in India.

NLMS Non Least Mean Square: Non Least Mean Square is a special form of LMS algorithm.

N-LOS Non-Line of Sight: Non-Line of Sight is radio transmission across a path that is partially obstructed, usually by a physical object in the innermost Fresnel zone.

NOC Network on Chip: Network on Chip is a communication subsystem on an integrated circuit (commonly called a “chip”), typically between IP cores in a system on a chip (SoC).

OFDM Orthogonal Frequency Division Multiplexing: Orthogonal Frequency Division Multiplexing is a method of encoding digital data on multiple carrier frequencies where a large number of closely spaced orthogonal sub-carrier signals are used to carry data.

OTA Operational Transconductance Amplifier: Operational Transconductance Amplifier is an amplifier whose differential input voltage produces an output current.

PFD Phase Frequency Detector: Phase Frequency Detector is an electronic circuit that compares phases of two signals.

PLL Phase Locked Loop: Phase Locked Loop is a control system that generates an output signal whose phase is related to the phase of an input signal.

PLP Perceptual Linear Predictive Analysis

PNN Probabilistic Neural Network: Probabilistic Neural Network is a feedforward neural network, which was derived from the Bayesian network and a statistical algorithm called Kernel Fisher discriminant analysis.

PSK Phase-Shift Keying: Phase-Shift Keying is a digital modulation scheme that conveys data by changing, or modulating, the phase of a reference signal.

PSO Particle Swarm Optimization: Particle Swarm Optimization is a computational method that optimizes a problem by iteratively trying to improve a candidate solution with regard to a given measure of quality.

QAM Quadrature Amplitude Modulation: Quadrature Amplitude Modulation is both an analog and a digital modulation scheme which conveys two analog message signals, or two digital bit streams, by changing the amplitudes of two carrier waves, using the amplitude-shift keying (ASK) digital modulation scheme oramplitude modulation (AM) analog modulation scheme.

QoS Quality of Service: Quality of Service refers to several related aspects of telephony and computer networks that allow the transport of traffic with special requirements.

QPSK Quadrate Phase Shift Keying Quadrature: Quadrate Phase Shift Keying Quadrature is a form of Phase Shift Keying in which two bits are modulated at once, selecting one of four possible carrier phase shifts (0, 90, 180, or 270 degrees).

RBI Reserve Bank of India: Reserve Bank of India is India's central banking institution, which controls the monetary policy of the Indian rupee.

RLS Recursive Least Square is an algorithm used in adaptive filter which recursively finds the filter coefficients that minimize a weighted linear least squares cost function relating to the input signals.

RMSE Root Mean Square Error: Root Mean Square Error is a frequently used measure of the differences between values predicted by a model or an estimator and the values actually observed.

RNN Recurrent Neural Network: Recurrent Neural Network is a class of neural network where connections between units form a directed cycle which creates an internal state of the network which allows it to exhibit dynamic temporal behavior.

ROI Region of Interest: Region of Interest is a selected subset of samples within a dataset identified for a particular purpose.

RTRL Real-Time Recurrent Learning: Real-Time Recurrent Learning is a training method with RNN.

RTS Reactive Tabu Search: Reactive Tabu Search is a Metaheuristic and a Global Optimization algorithm and is an extension of Tabu Search and the basis for a field of reactive techniques called Reactive Local Search and more broadly the field of Reactive Search Optimization.

SBP Systolic Blood Pressure: Systolic Blood Pressure is peak pressure in the arteries, which occurs near the end of the cardiac cycle when the ventricles are contracting.

SC Selection Combining: Selection Combining is a form of diversity method.

SIMO Single-Inductor Multiple-Output: Single-Inductor Multiple-Output is the use of multiple antennas at the transmitter and single antenna at the receiver.

SINR Signal to Interference and Noise Ratio: Signal to Interference and Noise Ratio is the ratio of signal power to noise power plus interference power.

SISO	Single-Input-Single-Output: Single-Input-Single-Output is the use of single antenna at both the transmitter and receiver.
SNR	Signal to Noise Ratio: Signal to Noise Ratio is the ratio of received signal power to the noise power.
SOC	System on Chip: System on Chip is an integrated circuit (IC) that integrates all components of a computer or other electronic system into a single chip.
SOI	Silicon on Insulator: Silicon on Insulator technology refers to the use of a layered silicon-insulator-silicon substrate in place of conventional silicon substrates in semiconductor manufacturing, especially microelectronics, to reduce parasitic device capacitance, thereby improving performance.
SOM	Self-Organizing Map: Self-Organizing Map is a type of ANN that is trained using unsupervised learning to produce a low-dimensional (typically two-dimensional), discretized representation of the input space of the training samples, called a map.
SRAM	Static Random Access Memory: Static Random Access Memory is a type of semiconductor memory that uses bistable latching circuitry to store each bit.
SSE	Sum of Squares of Errors: Sum of Squares of Error is the sum of the squared differences between each observation and its group's mean.
STBC	Space Time Block Code: Space Time Block Code is a technique used in wireless communications to transmit multiple copies of a data stream across a number of antennas and to exploit the various received versions of the data to improve the reliability of data-transfer.
STC	Space Time Codes: Space Time Codes is a method employed to improve the reliability of data transmission in wireless communication systems using multiple transmit antennas.
STTC	Space Time Trellis Code: Space Time Trellis Codes are a type of space-time code used in multiple-antenna wireless communications. This scheme transmits multiple, redundant copies of a trellis (or convolutional) code distributed over time and a number of antennas ('space').
STTuC	Space Time Turbo Code: Space Time Turbo Code is a special form of STTC.
SUR	Speech Understanding Research: Speech Understanding Research is a program founded by Advanced Research Projects Agency (ARPA) of the U.S. Department of Defense during 1970s.
S-V	Saleh-Valenzuela: Saleh-Valenzuela is a statistical model whose basic assumption is that multipath components (MPCs) arrive in clusters, formed by the multiple reflections from the objects in the vicinity of receiver and transmitter.

SVD Singular Value Decomposition: Singular Value Decomposition is a factorization of a real or complex matrix, with many useful applications in signal processing and statistics.

SVM Support Vector Machine: Support Vector Machine are supervised learning models with associated learning algorithms that analyze data and recognize patterns, used for classification and regression analysis.

TDNN Time Delay Neural Network: Time Delay Neural Network is an ANN architecture whose primary purpose is to work on continuous data.

TI Texas Instrument: Texas Instruments is an American company that designs and makes semiconductors, which it sells to electronics designers and manufacturers globally.

TS Takagi Sugeno: Takagi Sugeno is a type of fuzzy inference system.

TSK Takagi Sugeno Kang: Takagi Sugeno Kang is a modification of LS fuzzy inference system.

UWB Ultra Wide Band: Ultra Wide Band is a radio technology pioneered by Robert A. Scholtz and others which may be used at a very low energy level for short-range, high-bandwidth communications using a large portion of the radio spectrum.

VLIW Very Long Instruction Word: Very Long Instruction Word refers to a processor architecture designed to take advantage of instruction level parallelism.

VLSI Very Large Scale Integration: Very Large Scale Integration is the process of creating integrated circuits by combining thousands of transistors into a single chip.

WPD Wavelet Packet Decomposition: Wavelet Packet Decomposition is a wavelet transform where the discrete-time (sampled) signal is passed through more filters than the discrete wavelet transform (DWT).

WT Wavelet Transform: Wavelet Transform is a form of spectral transformation method used for analysis of signals with time and frequency content.

Part I
Intelligent Applications in Communication

This section includes five contributions which represent recent advances in intelligent application in communication and related areas. The work, ANFIS-based Symbol Recovery in Multi-Antenna Stochastic Channels, by Das et al. describes the application of fuzzy-based systems for symbol recovery in stochastic wireless channels. In the next chapter by Saikia et al., a method of ANN-based STBC decoding has been described. Purkayastha et al. describes the recovery of carrier phase in wireless channel using DPLL, which is described in the third chapter. Other works in this section include works by Choudhury et al. and Thakuria et al.

Chapter 1
ANFIS-Based Symbol Recovery in Multi-antenna Stochastic Channels

Banti Das, Manasjyoti Bhuyan and Kandarpa Kumar Sarma

Abstract Since stochastic wireless channels are highly random, fuzzy-based system are suitable options to deal with such uncertainty. This is because of the fact that the fuzzy system provides expert-level decision while tracking microscopic changes. Fuzzy system, however, requires support from artificial neural network (ANN)s for implementing inference rules. When fuzzy and ANN systems are combined, either neuro-fuzzy (NF) or fuzzy-neural (FN) frameworks are derived. Here, we propose an NF-based model for data recovery in multi-antenna setups when transmitted through stochastic wireless channels. Experimental results show that the proposed approach is computationally efficient.

Keywords Adaptive neuro-fuzzy inference systems (ANFIS) · Neuro-fuzzy systems (NFS) · Multiple-input multiple-output (MIMO) technology · Symbol recovery · Stochastic wireless channel

1.1 Introduction

The demand for greater data rate and higher bandwidth have necessitated the design and development of innovative solutions for ever-expanding mobile networks. One of the primary challenges in high data rate transmissions are due to the fading effects generated by the randomness present in the propagation medium. Further, the performance of such frameworks are limited by phenomena like co-channel interference (CCI) and severe fading due to vehicular movements and fast changing transmit–receive conditions. Channel fading and CCI contribute toward performance

B. Das (✉) · M. Bhuyan · K.K. Sarma
Department of Electronics and Communication Technology,
Gauhati University, Guwahati 781014, Assam, India
e-mail: bantids85@gmail.com

M. Bhuyan
e-mail: manasjyoti.b@gmail.com

K.K. Sarma
e-mail: kandarpaks@gmail.com

K.K. Sarma et al. (eds.), *Recent Trends in Intelligent and Emerging Systems*,
Signals and Communication Technology, DOI 10.1007/978-81-322-2407-5_1

degradation and renders reliable high data transmissions ineffective. The most effective technique to deal with fading is the exploitation of diversity [1].

Spatial diversity, also referred to as transmit and/or receive antenna diversity, represents a powerful means for combating the deleterious effects of fading [1]. Multiple-input multiple-output (MIMO) technology has received attention in wireless communication to increase data throughputs, link range, and transmission reliability without additional bandwidth or transmit power because it combines the benefits of transmit and receiver diversity. However, a very common form of uncertainty and stochastic behavior is observed in case of MIMO wireless channels. The associated time-dependent statistical variation in the channel properties coupled with high-speed transmission and CCI makes MIMO channel modeling an area filled with complexity and uncertainty. Soft-computing components such as fuzzy and neural systems are applied as alternate methods of statistical techniques used to model such uncertainties. In channel modeling, fuzzy systems are applicable where expert-level decision making is essential while artificial neural network (ANN) is applied comfortably for model-free approximations [2] with available process data. Therefore, while ANNs show numeric-quantitative capability, fuzzy systems exhibit symbolic capacity [2, 3]. Thus, hybrid systems formed by combinations of fuzzy and ANN methods have adaptability, parallelism, nonlinear processing, robustness, and learning in data-rich environment and excel in modeling uncertainty. The combination of fuzzy and ANN produces neuro-fuzzy system (NFS) or fuzzy-neural system (FNS). When fuzzy and ANN systems are combined, the composite system acquires the ability to provide numeric-qualitative, expert-level decision making and demonstrates greater adaptability and robustness while handling unknown processes or situations. Such attributes of FNS and NFS make them suitable for modeling MIMO channels and related applications.

With the dynamic capabilities, fuzzy-based systems have already been considered for wireless domain starting from the use of fuzzy adaptive filters for nonlinear channel equalization in [4] to applications in data-rich environments. In [4], adaptive filters adopted were namely the recursive least square (RLS) and the least mean square (LMS) adaptation algorithm and used them together with fuzzy systems. Fuzzy-based channel estimation approach is evaluated for the multipath fading CDMA channel in [5]. A takagi-sugeno-kang (TSK) fuzzy model for channel estimation in orthogonal frequency division multiplexing (OFDM) system is proposed in [6]. This work is extended subsequently for MIMO-OFDM systems [7]. Most of the reported works do not focus on the symbol recovery part of multi-antenna setups used in stochastic MIMO wireless channels.

Here, we propose a NFS-based model for data recovery in multi-antenna setups when transmitted through stochastic wireless channels. We specifically design a Takagi–Sugeno (TS)-based adaptive neuro-fuzzy inference systems (ANFIS) for symbol recovery in MIMO channel with CCI and severe fading. We design the ANFIS to implement certain inference rules which enable satisfactory recovery of block-sized data in MIMO channels. Experimental results show that the proposed approach is computationally efficient and is suitable for adaptive receiver designs.

The rest of the paper is organized as below:

Section 1.2 provides a brief theoretical background. The proposed system model is described in Sect. 1.3. Section 1.4 includes the results and discussion. The work is concluded in Sect. 1.5.

1.2 Theoretical Consideration

Here, we briefly describe about the basic theoretical notions related to the work.

Section 1.2.1 discusses about ANN. Fuzzy systems is briefly covered in Sect. 1.2.2. Some basic considerations of NFS is included in Sect. 1.2.3.

1.2.1 ANN

An ANN is a mathematical tool or computational model based on the analogy of biological nervous systems [3] consisting of interconnected group of artificial neurons and information processing using a connectionist approach. An ANN may be single or multilayered. The knowledge gained during the training phase is stored in the interconnecting neurons and used subsequently.

1.2.2 Fuzzy Systems

System modeling based on the conventional mathematical tools is not well suited for dealing with ill-defined and uncertain systems. The invention or proposition of fuzzy sets was done by Zadeh in 1965 [8] as a means to capture and represent the real world with its fuzzy data due to its uncertainty. Fuzzy logic starts with the concept of fuzzy set, a set without a crisp, clearly defined boundary. The idea of grade of membership is the backbone of fuzzy set theory.

The primary components of a fuzzy-based system or fuzzy inference system are fuzzification, rule base, inference, and defuzzification stages as shown in Fig. 1.1. There are two types of fuzzy inference systems that can be implemented for various applications. These are Mamdani-type and Sugeno-type.

Mamdani's fuzzy inference method [9] is an attempt to control a steam engine and boiler combination by synthesizing a set of linguistic control rules obtained from experienced human operators. It is the most commonly used fuzzy methodology. Mamdani's inference method expects the output membership functions (MF) to be fuzzy sets. After the aggregation process, there is a fuzzy set for each output variable that needs defuzzification.

Sugeno method of fuzzy inference introduced in 1985 [10], is similar to the Mamdani method in many respects. The first two parts of the fuzzy inference process,

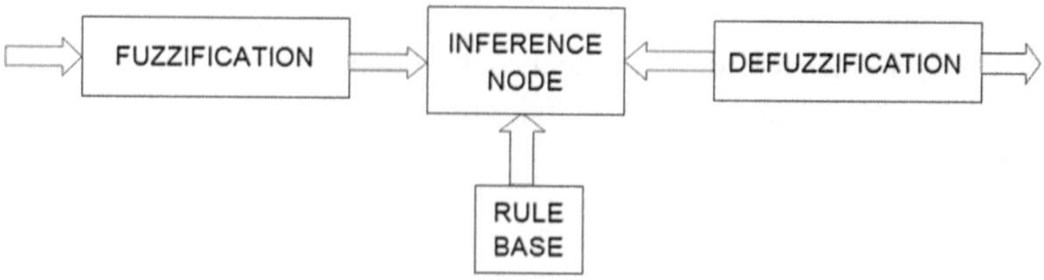

Fig. 1.1 Components of a fuzzy system

fuzzifying the inputs, and applying the fuzzy operator, are exactly the same. The main difference between Mamdani and Sugeno is that the Sugeno output MFs are either linear or constant.

1.2.3 Neuro-Fuzzy Systems (NFS)

As already described, the fuzzy systems provide expert-level decision making by following microscopic variation. ANNs are able to work with model-free data, learn from the environment, retain the knowledge, and use it subsequently. Thus, fuzzy and ANN systems having the symbolic qualitative capacity and numeric-quantitative capability, respectively, can be applied together to deal with situations where higher precision of estimation is crucial. Though NFS and FNS combinations are available, we have used NFS-based approach.

A noteworthy contribution in the NF domain is the formulation of ANFIS (adaptive neuro-fuzzy inference system)-a system developed by J. Roger Jang which has found numerous applications in a variety of fields. Here, the fuzzy rules can either be based on the Mamdani method or Sugeno [11] method. However, Sugeno is a more compact and computationally efficient representation than a Mamdani system. The Sugeno system lends itself to the use of adaptive techniques for constructing fuzzy models. So, here Sugeno model is used.

Let the considered fuzzy inference system (FIS) has two inputs x and y and one output f. For a first-order Sugeno fuzzy model, a common rule set with two fuzzy emphif–*then* rules is as follows:

Rule 1 : If x is A_1 and y is B_1, then $f_1 = p_1x + q_1y + r_1$
Rule 2 : If x is A_2 and y is B_2, then $f_2 = p_2x + q_2y + r_2$

For the Sugeno model, the type-3 fuzzy reasoning mechanism is given in Fig. 1.2. The type-3 fuzzy reasoning is preferred for modeling both linear and nonlinear system because of the linear dependence of each rule on the input variables. The corresponding equivalent ANFIS architecture is shown in Fig. 1.3 [12]. Here, the node functions in the same layer are of the same function family as described below.

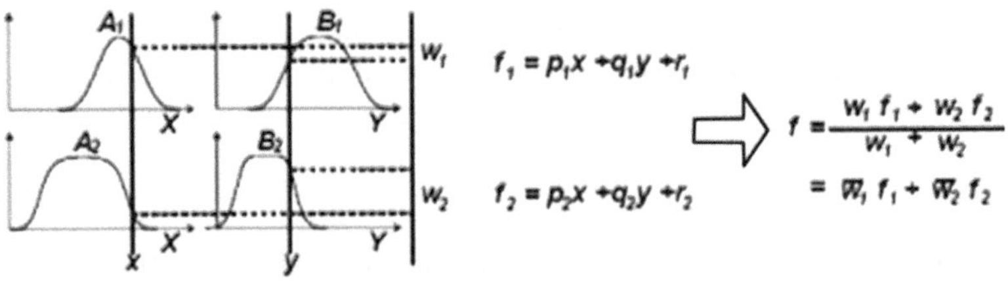

Fig. 1.2 TS type-3 fuzzy reasoning

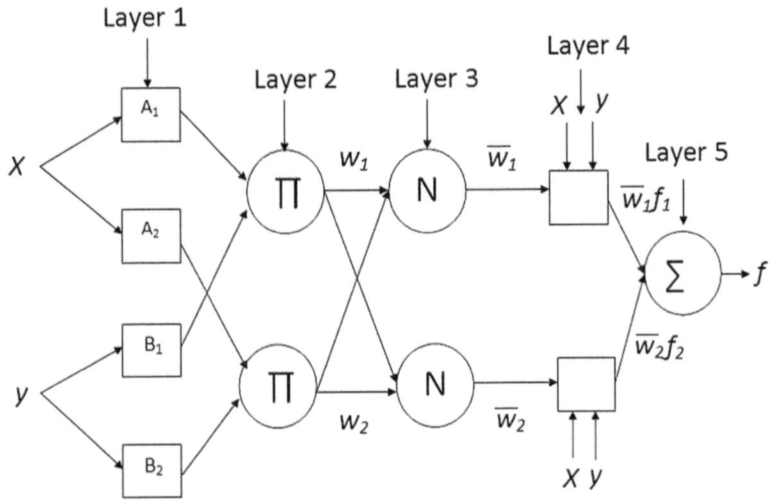

Fig. 1.3 Equivalent ANFIS

Layer 1: Every node i in this layer is a square node with a node function

$$O_i^1 = \mu_{A_i}(x) \tag{1.1}$$

where x is the input to node i, and A_i is the linguistic label (small, large etc.) associated with this node function. In other words, O_i^1 is the MF of A_i and it specifies the degree to which the given x satisfies the quantifier A_i. Parameters in this layer are referred to as premise parameters.

Layer 2: Every node in this layer is a circle node labeled $\prod$ which multiplies the incoming signals and sends out the product. For instance,

$$w_i = \mu_{A_i}(x) \times \mu_{B_i}(y) \tag{1.2}$$

Each node output represents the firing strength of a rule. Here x and y are the inputs to node i and A_i, B_i are the linguistic labels associated with MFs $\mu_{A_i}(x)$ and $\mu_{B_i}(y)$, respectively

Layer 3: Every node in this layer is a circle node labeled N. The ith node calculates the ratio of the ith rules firing strength to the sum of all rules' firing strengths:

$$\overline{w_i} = \frac{w_i}{w_1 + w_2}, i = 1, 2. \tag{1.3}$$

Output of this layer, $\overline{w_i}$ will be called the normalized firing strengths. Here, w_1, w_2 are the firing strengths to the rules 1 and 2.

Layer 4: Every node i in this layer is a square node with a node function

$$O_i^4 = \overline{w_i} f_i = \overline{w_i}(p_i x + q_i y + r_i) \tag{1.4}$$

where $\overline{w_i}$ is the output of layer 3, and p_i, q_i, r_i are the parameter set. Parameters in this layer will be referred to as consequent parameters.

Layer 5: The single node in this layer is a circle node labeled $\sum$ that computes the overall output as the summation of all incoming signals, i.e.,

$$O_1^5 = \text{overall output} = \sum_i \overline{w_i} f_i = \frac{\sum_i w_i f_i}{\sum_i w_i} \tag{1.5}$$

Thus, ANFIS uses a hybrid learning procedure for estimation of the premise and consequent parameters [12].

1.3 Proposed System of Symbol Recovery Using ANFIS-Based Approach

Here, we describe the proposed ANFIS-based approach for recovery of symbols in multi-antenna setups in stochastic wireless channels. The formulated problem may be depicted as in Fig. 1.4. Using a QPSK-modulated signal for a 2×2 MIMO setup the input-output relation may be written as

$$y_i(k + 1) = [x_i(k) + x_i(k - \tau)]H(i, k) + n \tag{1.6}$$

$$y_i(k + n) = F[x_i(k, \tau), x_i(k)H(i, k)] + n \tag{1.7}$$

where $F(.)$ is a mapping function representing the transformation in the transmission, τ is the delay associated with multipath fading. In another form

$$y(n) = FM[y(n - 1), \ldots, x(n - 1), \ldots, [H], \ldots] \tag{1.8}$$

where FM is a fuzzy mapping generated by the first layer of ANFIS. The requirement is to design another fuzzy-supported ANN-based process such that its output is given by

$$yG(n) = FM_G[y(n - 1), \ldots, x(n - 1), \ldots, [H], [W], [V], \theta] \tag{1.9}$$

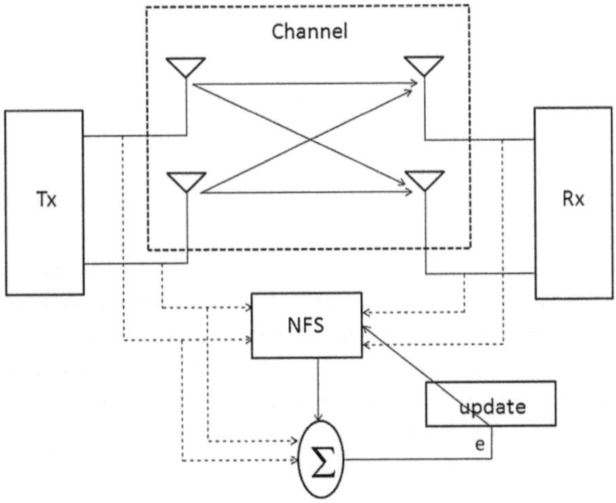

Fig. 1.4 System block diagram

so that $yG(n) \rightarrow y(n)$ in terms of a cost function. Here, $[W]$, $[V]$, and θ are forward and backward connectionist weights and biases, respectively. Therefore, the training of a fuzzy process is related to the minimization of a cost function expressed as

$$CF = \frac{1}{TN \times VD} \sum \sum d(y_{di}, y_{ai}) \tag{1.10}$$

where TN is the number of training samples, VD is the dimension of samples, d(.) is a distance measure, y_{di} is the desired output and y_{ai} is the actual output [13]. The channel matrix **H** represents the wireless medium through which the propagation takes place. The channel is considered to be a time variant system with an impulse response given by [14]:

$$h(t, \tau) = \sum_{g=0}^{N_p - 1} \alpha_g(t, \tau) \exp(j(2\pi f_c \tau_g(t) + \phi(t, \tau)))\delta(\tau - \tau_g(t)); \tag{1.11}$$

where N_p is the number of multipath components, $\alpha_g(t, \tau)$ is the amplitude components, and $\tau_g(t)$ is the excess delay component caused by the gth multipath component at time t and δ is delta function. The inverse dynamics allows a definition

$$[H] = G[y_i(k), x_i(k), x_i(k - \tau)]. \tag{1.12}$$

The channel matrix H can be determined from the inverse dynamics G obtained from the following sets of data:
$[y_i(k), x_i(k), x_i(k - \tau)]$.

The following two sections describe the experimental considerations related to the formulation of NFS approach of symbol recovery in multi-antenna systems.

1.3.1 Input Conditioning

The multi-antenna setup represented by the MIMO channel input-output relationship is expressed as

$$X_n = H(n)s(n) + v(n) \tag{1.13}$$

where x is an $M \times 1$ vector with $x_i(n), i = 1, 2, \ldots, M$ as the elements, $s(n)$ representing the signal symbols, $v(n)$ denotes additive background noise with $H(n)$ being the $M \times N$ channel matrix which is normally Rayleigh multipath fading. The respective channel sets are generated and the signals are combined following time and frequency domain considerations where there are convolution and multiplication processes, respectively, between -10 and 10 dB. These are normalized and confined by the following set of norms:

$$f(x) = NL, -0.66 \leq x < -0.99;$$

$$= NM, -0.33 \leq x < -0.66 \tag{1.14}$$

$$= NS, 0 < x < -0.33; \tag{1.15}$$

$$= CS, x = 0; \tag{1.16}$$

$$= PL, 0.66 \leq x \leq 0.99; \tag{1.17}$$

$$= PM, 0.33 \leq x \leq 0.66; \tag{1.18}$$

$$= PS, 0 \leq x < 0.33; \tag{1.19}$$

The inputs are divided into in-phase and quadrature components to allow the NFS to learn the individual signal segments separately. The decoupled components of the signals allow the NFS to sustain the learning through the complete range of variations observed in the signals while propagating through a time-varying MIMO channel.

1.3.2 Fuzzification

One of the primary blocks of our proposed approach is a fuzzy equalizer. In our ANFIS-based channel equalizer, we have used Gaussian and Bell MFs. The bell-shaped function can approach a nonfuzzy set if their parameters are properly tuned. It has the advantage of being smooth and nonzero at all points. Further, the Bell MF is invariant to Fourier transform etc. The generalized Bell function depends on three parameters a, b, and c and is expressed as

$$bell(x; a, b, c) = \frac{1}{1 + |\frac{x-c}{a}|^{2b}} \tag{1.20}$$

where c determines the center of the curve or MF, a is the half width, and b (together with a) controls the slopes at the crossover points (where MF value is 0.5) and the

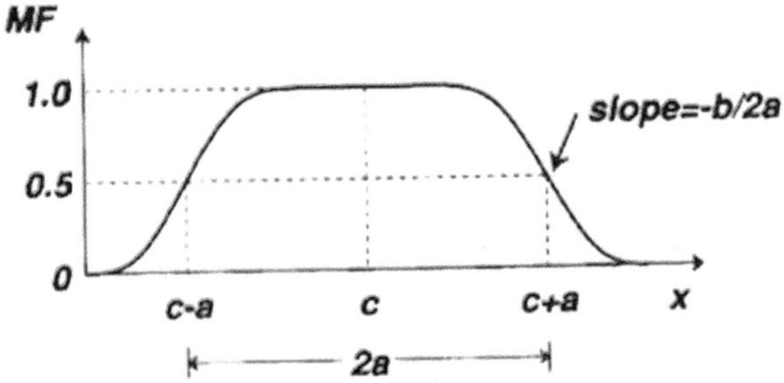

Fig. 1.5 Physical meanings of the parameters in the bell membership function

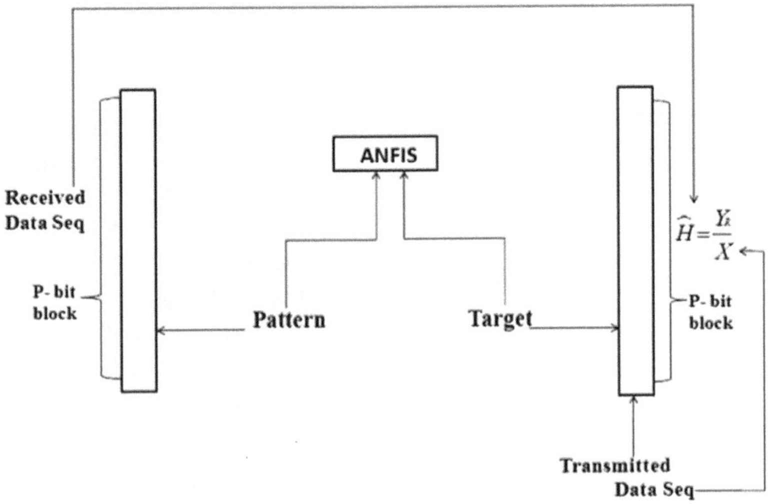

Fig. 1.6 Static setup for SISO and MIMO configuration

parameter b is usually positive. A desired generalized bell MF can be obtained by a proper selection of the parameter set a, b, c. Specifically, we can adjust c and a to vary the center and width of the MF, and then use b to control the slopes at the crossover points. We have also considered the Gaussian membership function because it closely approximates the randomness observed in wireless channels. A detailed set of experiments are carried out. Respective performances observed derived by considering these two MFs and their impact in ascertaining the system performance (Fig. 1.5).

1.4 Result and Discussion

The samples are accumulated as block size (as P-bit groups) for the SISO and 2×2 MIMO cases with AWGN values between 0 and 12 dB. The basic setup for SISO and MIMO configuration is shown in Figs. 1.6 and 1.7. Some of the parameters used

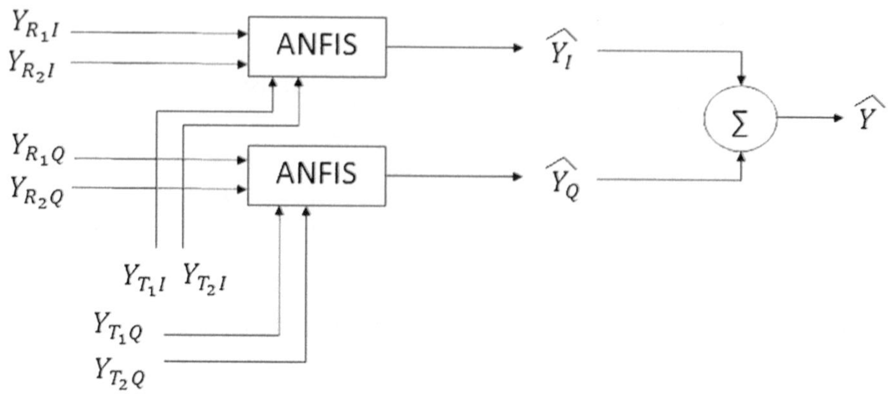

Fig. 1.7 Real-imaginary part

Table 1.1 Parameters used for simulating channel using Clarke-Gans model

Sl. no.	Parameter	Value
1	Freq., f_c	900 MHz
2	Mobile speed, V	3–100 kmph
3	No. of paths	8
4	Wavelength, λ	3×10^8
5	Doppler shift, f_m	$\frac{V}{\lambda}$
6	Sampling Freq., f_s	$8\times f_m$
7	No. of samples, N	6400
9	Sampling period, T_s	$\frac{1}{f_s}$
10	Antenna configuration	2×2

to perform the experiments are summarized in Table 1.1 [15]. The setup shown in Fig. 1.6, represents a static approach where pattern and target data vectors are presented as columns to the ANFIS. It takes longer training time and is computationally inefficient. Further, it requires more resources. With increase in the size of the data blocks, it results in slowing down. Hence, we formulated a dynamic approach which is shown in Fig. 1.8. This method is dynamic because depending upon the antenna configuration, the column vectors can be arranged. The training and testing is however carried out using the approach shown in Fig. 1.6 and the process is carried out in decoupled form.

ANFIS parameters are adjusted so as to reduce the error measure defined by the sum of the square of the difference between the actual and desired output. The root mean square error (RMSE) is calculated using

$$\text{RMSE} = \sqrt{\frac{1}{N}\sum_{t=1}^{N}(A_t - F_t)^2} \tag{1.21}$$

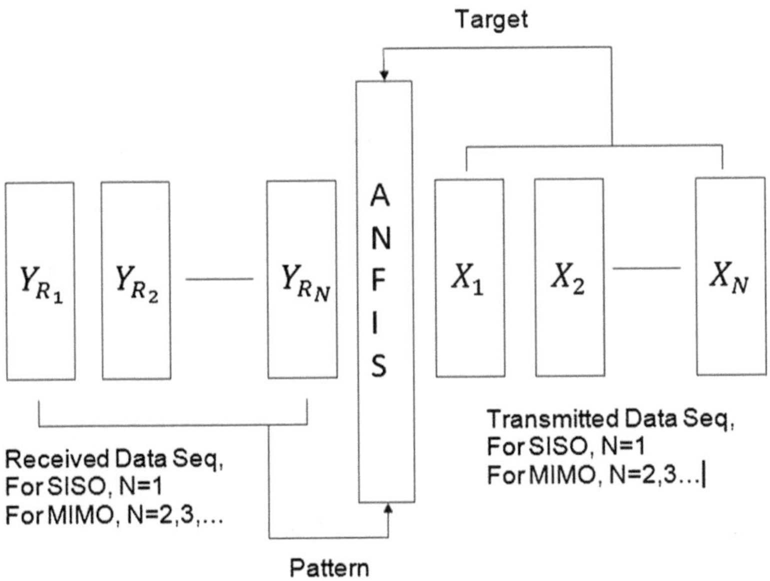

Fig. 1.8 Dynamic setup for SISO and MIMO configuration for training the ANFIS

Table 1.2 Number of iterations required compared to statistical and ANN-based approaches [15]

Sl. no.	Method	Iterations
1	LS	52
2	MMSE	33
3	MLP	94
4	3L-FF	23
5	Temporal-MLP	23
6	Proposed	17

where A_t and F_t are actual and fitted values, respectively, and N is the number of training or testing sample. The RMSE is used as the primary criterion to ascertain the extent of learning acquired by the ANFIS.

For SISO case, ANFIS is employed for both the Rayleigh flat-faded and frequency-selective channels. For SISO flat fading channel, we have used the BPSK method. The convergence of RMSE and time required due to change in MF are shown in Tables 1.2 and 1.3. Again, we have also considered the performance for both the Gaussian MF and Bell-shaped MF. Here, we see that the best average RMSE is obtained for Bell-shaped MF with 10 such functions with an average RMSE of 2×10^{-5} obtained within 1 s. Similarly, for the SISO frequency-selective fading case, the performance is evaluated for a range of MF values from 2 to 10. From Table 1.2, it is observed that for Gaussian MF of 8, the best performance is obtained with the average RMSE

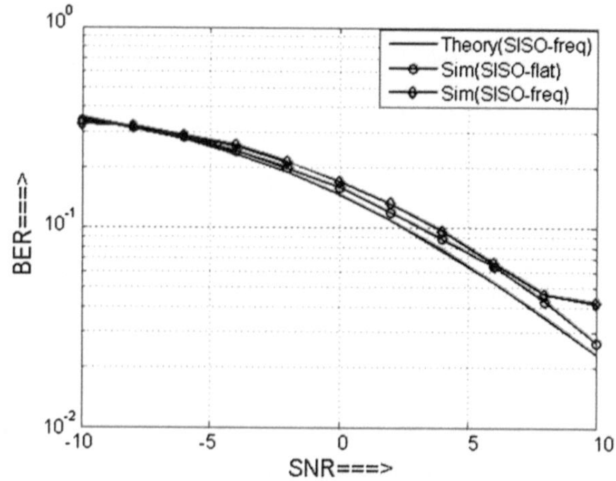

Fig. 1.9 SISO flat fading and frequency-selective fading case

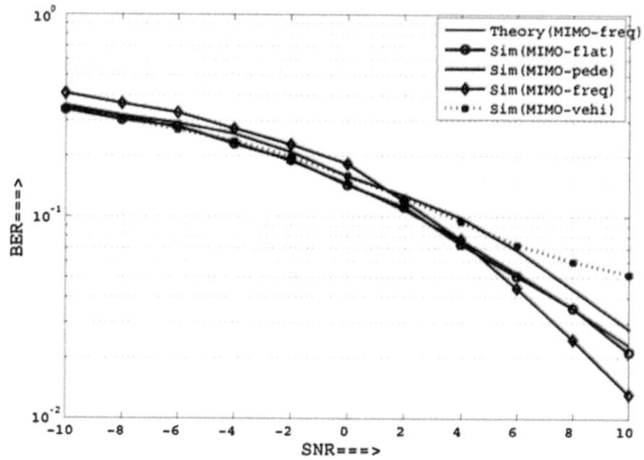

Fig. 1.10 MIMO flat fading and frequency-selective fading case

of 0.00011 which is obtained with one epoch completed below 1.5 s consistent with at least ten trials. Similarly, for Bell-shaped MF of 10, the best RMSE of 0.00091 is obtained within an epoch under 1.5 s.

The bit error rate (BER) curve for SISO flat fading and SISO frequency selective for a range of SNR between −10:10 dB is shown in Fig. 1.9. For the MIMO 2×2 channel, both the single path link MIMO and multiple path link MIMO is considered. Moreover, ITU-MIMO pedestrian and ITU-MIMO Vehicular cases are also considered. The MIMO fading channel with frequency-selective multipath link models, each discrete path act as an independent Rayleigh fading process. CCI sources are considered to be present as well. The BER curve for MIMO theoretical and simulated SNR range −10:10 dB is shown in Fig. 1.10. From the Tables 1.2 and 1.3, we can say that for the MIMO case, no definite statistical regularity is observed for both

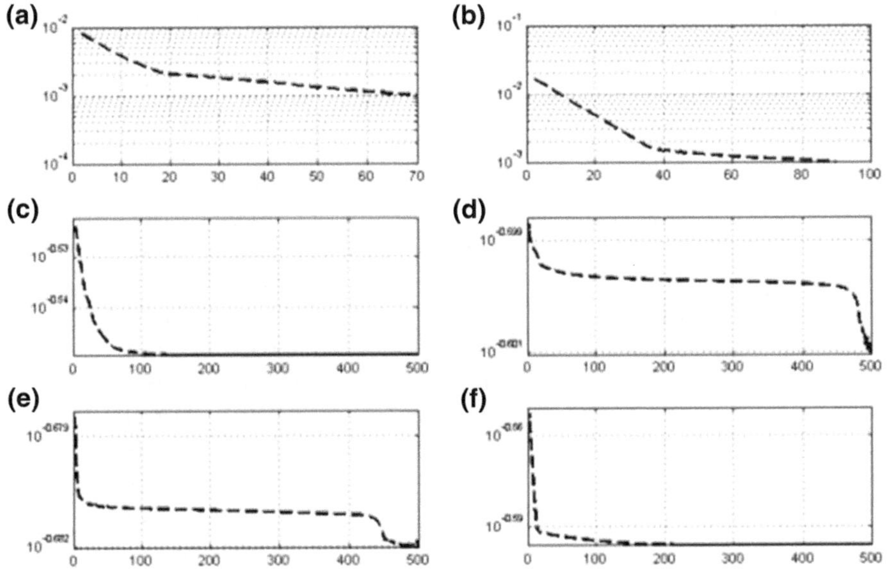

Fig. 1.11 RMSE convergence plots for **a** SISO frequency flat **b** SISO frequency-selective **c** MIMO frequency flat **d** MIMO frequency selective **e** MIMO pedestrian **f** MIMO vehicular

the Gaussian and Bell-shaped MFs for all of the channel states considered. This is perhaps due to the stochastic nature of the MIMO channel. The RMSE curves for all the specific cases mentioned is shown in Fig. 1.11.

The situation is somewhat regular while adopting the dynamic setup shown in Fig. 1.8.

The results obtained are compared with [13, 15, 16]. The computational complexity associated with the proposed approach is summarized in Table 1.4 involving 6400 b blocks transmitted through severely faded MIMO channels. The ANFIS generates comparable BER values but computation speed is significantly better. The ANFIS shows at least 50 % increase in computational speed compared to least square (LS), minimum mean square error (MMSE), multi-layer perceptron (MLP), three-layered feedforward ANN (3L-FF), and recurrent neural network (RNN)-based approaches reported in [15] which establishes its usefulness. Further, the RMSE convergence is completed in less number of epochs in presence of severe fading and CCI meaning better learning. Thus, the proposed approach is effective and is suitable for the design of adaptive receivers for data-rich environments.

Table 1.3 Calculation of average RMSE for 500 epochs for different channel conditions

Case	Channel type	No. of MFs	Average RMSE for 500 epochs	
			Gaussmf	Gbellmf
1	SISO flat fading	2	0.1633	0.0097
		4	0.0153	0.0069
		6	0.0069	0.0023
		8	0.0059	0.0034
		10	0.0039	0.00002
2	SISO Freq. selective	2	0.0132	0.0159
		4	0.0050	0.0035
		6	0.0037	0.0021
		8	0.00011	0.0370
		10	0.0027	0.00091
3	MIMO flat fading (2 × 2)	2	0.4019	0.2841
		4	0.2642	0.2830
		6	0.3699	0.2923
		8	0.3893	0.2833
		10	0.4341	0.3188
4	MIMO Freq. selective (2 × 2)	2	0.3939	0.4129
		4	0.3444	0.3917
		6	0.4191	0.3364
		8	0.3740	0.2991
		10	0.4112	0.2513
5	MIMO pedestrian (2 × 2)	2	0.4371	0.3173
		4	0.2568	0.3202
		6	0.3700	0.2084
		8	0.2409	0.3596
		10	0.1315	0.2511
6	MIMO vehicular (2 × 2)	2	0.4041	0.2542
		4	0.3538	0.3375
		6	0.3300	0.3528
		8	0.3952	0.3057
		10	0.3689	0.3335

Table 1.4 Time taken for 500 epochs for different channel conditions

Case	Channel type	No. of MFs	Time taken for 500 epochs (in s)	
			Gaussmf	Gbellmf
1	SISO flat fading	2	0.87	0.85
		4	0.86	0.85
		6	0.84	0.88
		8	0.87	0.88
		10	0.88	0.90
2	SISO Freq. selective	2	1.50	0.99
		4	1.52	1.57
		6	1.56	1.52
		8	1.58	1.53
		10	1.56	1.54
3	MIMO flat fading (2 × 2)	2	1.60	1.64
		4	1.59	1.64
		6	1.68	1.66
		8	1.68	1.70
		10	1.74	1.83
4	MIMO Freq. selective (2 × 2)	2	1.60	1.61
		4	1.61	1.70
		6	1.63	1.75
		8	1.71	1.81
		10	1.66	1.72
5	MIMO pedestrian (2 × 2)	2	1.63	1.64
		4	1.63	1.65
		6	1.60	1.67
		8	1.64	1.73
		10	1.64	1.68
6	MIMO vehicular (2 × 2)	2	1.83	1.62
		4	1.83	1.64
		6	1.79	1.74
		8	1.64	1.69
		10	1.70	1.77

1.5 Conclusion

Here, we proposed an ANFIS-based symbol recovery system for multi-antenna setups used in stochastic wireless channels. We configured the ANFIS in split form using both Gaussian and Bell MFs for SISO and MIMO setups with flat and frequency-selective behavior. We further performed trials using ITU-pedestrian and vehicular channels. The experimental results show that the proposed approach is at least 50 % more efficient compared to ANN-based approaches reported earlier.

References

1. Duman TM, Ghrayeb A (2007) Coding for MIMO communication systems. Wiley, England
2. Ross TJ (2008) Fuzzy logic with engineering applications, 2nd edn. Wiley India, New Delhi
3. Haykin S (2003) Neural networks-a comprehensive foundation, 2nd edn. Pearson Education, New Delhi
4. Wang LX, Mendel Jerry M (1993) Fuzzy adaptive filters with application to nonlinear channel equalization. IEEE Trans Fuzzy Syst 1(3):161–170
5. Niemi A, Joutsensalo J, Ristaniemi T (2000) Fuzzy channel estimation in multipath fading CDMA channel. The 11th IEEE international symposium on personal. Indoor Mobile Radio Commun 2:1131–1135
6. Zhang J, He ZM, Wang XG, Huang YY (2006) A TSK fuzzy approach to channel estimation for OFDM systems. J Electron Sci Technol China 4(2)
7. Zhang J, He ZM, Wang XG, Huang YY (2007) TSK fuzzy approach to channel estimation for MIMO-OFDM systems. IEEE Signal Proc Lett 14(6):381–384
8. Zadeh LA (1965) Fuzzy sets. Inf Control 8:338–353
9. Mamdani EH, Assilian S (1975) An experiment in linguistic synthesis with a fuzzy logic controller. Int J Man Mach Stud 7(1):1–13
10. Sugeno M (1985) Industrial applications of fuzzy control. Elsevier Science Pub. Co, Amsterdam
11. Takagi T, Sugeno M (1983) Derivation of fuzzy control rules from human operator's control actions. In: Proceedings of the IFAC symposium fuzzy information, knowledge representation and decision analysis, pp 55–60
12. Jang RJ (1993) ANFIS: Adaptive-network-based fuzzy inference system. IEEE Trans Syst Man Cybern 23(2):665–685
13. Sarma KK, Mitra A (2012) MIMO channel modeling: suitability between neuro-fuzzy and fuzzy-neural approaches. National conference on computational intelligence and signal processing (CISP), pp 12–17
14. Molisch AF (2005) Wireless communications, 1st edn. John Wiley, Indian Reprint Systems, New Delhi
15. Sarma KK, Mitra A (2012) Estimation of MIMO wireless channels using artificial neural networks. Cross disciplinary applications of artificial intelligence and pattern recognition. doi:10.4018/978-1-61350-429-1.ch026
16. Gogoi P, Sarma KK (2012) Channel estimation technique for STBC coded MIMO system with multiple ANN blocks. Int J Comput Appl 50:10–14

Chapter 2
STBC Decoding with ANN in Wireless Communication

Samar Jyoti Saikia and Kandarpa Kumar Sarma

Abstract Diversity techniques can be used to reduce the ill effects of multipath fading observed in wireless channels. Multiple-input-multiple-output (MIMO) technology is a promising application of multiple antennas at both transmitter and receiver to improve communication performance by achieving spatial diversity. The concept of space–time coding has arisen from diversity techniques using smart antennas. With the implementation of data coding and signal processing at both sides of transmitter and receiver, space–time coding now is more effective than traditional diversity techniques. Space–time block codes (STBC) were designed to achieve the maximum diversity order for the given number of transmit and receive antennas subject to the constraint of having a simple linear decoding algorithm. Application of ANNs for STBC decoding is such an area which offers solutions to tackle the intricacies associated with the fluctuations observed in multipath propagation which is always the problem area in wireless communication.

Keywords Multi-input-multi-output (MIMO) system · Space–time block codes (STBC) · Artificial neural network (ANN) · Multipath fading · Rayleigh fading channels · Rician fading channels

2.1 Introduction

During the last decade, in wireless local area networks and cellular mobile systems, the demand for capacity has grown explosively. The need for wireless Internet access and multimedia applications requires an increase in information throughput with

S.J. Saikia (✉)
Department of Electronics and Communication Engineering,
Assam Don Bosco University, Guwahati, Assam, India
e-mail: samar.saikia@dbuniversity.ac.in

K.K. Sarma
Department of Electronics and Communication Technology,
Gauhati University, Guwahati 781014, Assam, India
e-mail: kandarpaks@gmail.com

K.K. Sarma et al. (eds.), *Recent Trends in Intelligent and Emerging Systems*,
Signals and Communication Technology, DOI 10.1007/978-81-322-2407-5_2

order of magnitude compared to the data rates made available by today's technology. The use of multiple antennas at both the transmitters and receivers in the system makes this increase in data rate possible. Space–time wireless technology that uses multiple antennas along with appropriate signaling and receiver techniques offers a powerful tool for improving wireless performance. In various wireless networks and cellular mobile standards, some aspects of this technology have already been incorporated. More advanced MIMO techniques are planned for future mobile networks, wireless local area network (LANs), and wide area network (WANs). Here, we discuss the use of STBC codes in MIMO systems implemented using artificial neural network (ANN). The primary reason behind the use of ANN for such a purpose is the fact the ANN is a learning-based system and can use the knowledge acquired for subsequent processing.

2.2 Background

Most work on wireless communications had focused on having an antenna array at only one end of the wireless link, usually at the receiver. Seminal papers by Foschini and Gans [1], Foschini [2], and Telatar [3] enlarged the scope of wireless communication possibilities by showing that when antenna arrays are used at both ends of a link, substantial capacity gains are enabled by the highly scattering environment. Many established communication systems use receive diversity at the base station. For example, global system for mobile communications (GSM) [4] base station typically has two receive antennas. This receive technology is used to improve the quality of the uplink from mobile to base station without adding any cost, size, or power consumption to the mobile [5].

2.2.1 Multi-Input-Multi-Output (MIMO) System

Multiple-input and multiple-output system contains multiple antennas at both transmitter and receiver. Figure 2.1 shows a MIMO system with TxM transmits antennas and RxN receive antennas. The received signal Y can be given by the following matrix equation:

$$Y = Hx + n \tag{2.1}$$

Here, x is the transmitted signal vector, n is the statistically independent complex zero-mean Gaussian random variables with equal variance [6], H is channel between transmitter and receiver which can be represented by matrix $T_{xM} \times R_{xN}$. The channel matrix is formed by h_{MN} which are gain coefficients modeling random phase shifts and channel gains.

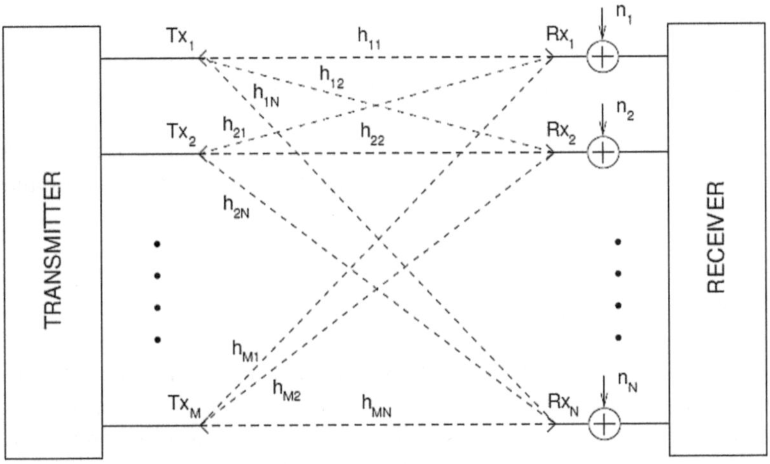

Fig. 2.1 MIMO system model

2.2.2 *Statistical Models for Wireless Channels*

In wireless communication, the received signal is the combination of many replicas of the original signal impinging on receiver from many different paths. The signals on these different paths can constructively or destructively interfere with each other. This is referred as multipath fading [7, 8].

2.2.2.1 Rayleigh Fading Channels

In Rayleigh fading [3], no line-of-sight (LOS) component is present. It is typically encountered in land mobile channels in urban areas where there are many obstacles which make LOS paths rare. This represents the worst fading case.

2.2.2.2 Rician Fading Channels

Wireless channels where a LOS component is present due to absence of high rise structures is modeled by the Rician fading. If a LOS path is present (or one path which dominates the rest), the Gaussian approximation usually preferred needs to be reconsidered.

2.2.2.3 Space–Time Coding (STC)

Space–time coding is used in wireless communication for transmitting multiple, redundant copies of a data stream to the receiver in the hope that at least some of them will survive the physical path between transmission and reception in a good state to allow reliable decoding. STC can be divided into three types. First space–time trellis code (STTC) which provides both coding and diversity gain [9]. The second type of STC is space–time turbo codes (STTuC), a combination of space–time coding and turbo coding [10]. The third type is space–time block codes (STBC) which provides diversity gain but not coding gain [10, 11]. So it is less complex than STTC and STTuC.

2.2.2.4 Artificial Neural Network (ANN)

An artificial neural network (ANN) is a parallel distributed processor that acquires knowledge through a learning (training) process. An artificial neuron is a simplistic representation that emulates the signal integration and threshold firing behavior of biological neurons by means of mathematical equations. The advantages of ANN are their nonlinearity, input–output mapping, adaptivity, evidential response, contextual information, fault tolerance, uniformity of analysis and design, neurobiological analogy, and diversity of types and topologies.

2.3 System Model

The proposed system model comprises of the blocks as shown in Fig. 2.2. A signal is generated using 10^6 b. The signal is sent to the modulator where it is modulated with different modulation schemes, then it is encoded with Alamouti STBC encoder and transmitted through the multipath fading channel. After the receiver has received the signal it is decoded in Alamouti decoder and passed to the demodulator and the desired signal is obtained and BER is compared between the transmitted signal and the received signal. The ANN in feedforward form trained with back propagation (BP) algorithm is used to perform the STBC decoding in the MIMO setup.

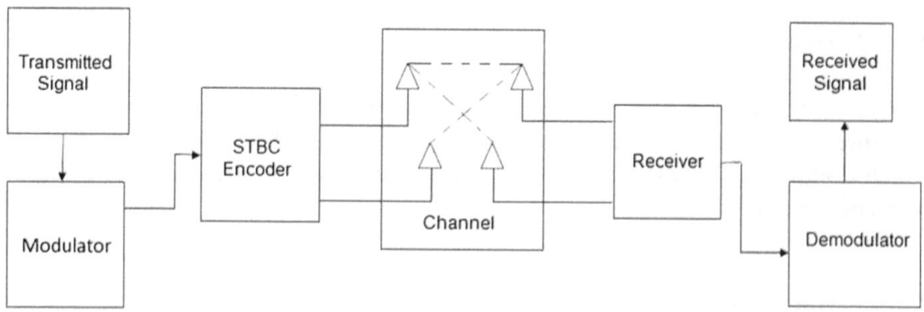

Fig. 2.2 System model

Table 2.1 Parameters used for simulating the Rayleigh channel

Channel type	Rayleigh channel
Input sample period	$1.0000e^{-005}$
Doppler spectrum	$[1 \times 1 Doppler.jakes]$
Max doppler shift	100 Hz
Path delays	$[01.500e^{-005}\ 3.2000e^{-005}]$
Average path gain in dB	$[0\ -3\ -3]$
Path gains	$[0.3814\ -\ 0.1735i\ -\ 0.0962\ -\ 0.4042i\ 0.1016\ -\ 0.1244i]$
Channel filter delay	4
Number of sample processes	100

2.3.1 Simulating the Rayleigh Channel

We have simulated a Rayleigh channel having three multipath components. Table 2.1 gives the parameters for simulating the Rayleigh channel.

2.3.2 Comparison Between Three Channels

Figure 2.3 shows BER versus Eb/No performance of the Alamouti scheme with one transmitter and one receiver with coherent BPSK modulation. Here, we can see that BER performance is poor. For -10 dB we get the BER of 10^{-1}, at 0 dB we get $10^{-1.5}$ and at 10 dB the BER is $10^{-2.5}$. Next, we have considered a system of two transmitters and one receiver in Fig. 2.4. Here theoretical and simulated BER performance of BPSK signals in Rayleigh fading channel is observed. We can see that the theoretical and simulated BER is almost the same and BER performance is better than one transmitter and one receiver system. At -10 dB we get BER of 10^{-1}, at 0 dB BER of $10^{-2.5}$ is recorded, and about 10 dB the BER performance is about $10^{-4.5}$.

The BER performance of BPSK modulated signal in AWGN, Rayleigh fading channel, and in MIMO channel is compared. As expected, Fig. 2.5 shows that the BER performance of the BPSK signal with Alamouti STBC with two transmitters and two receivers in Rayleigh fading channel is better than the signal without STBC. Table 2.2 gives the BER comparison of the three channels.

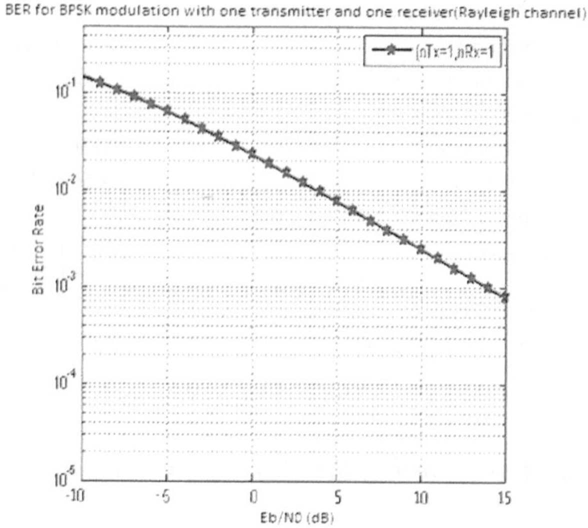

Fig. 2.3 BER for BPSK modulation with one transmitter and one receiver (Rayleigh channel)

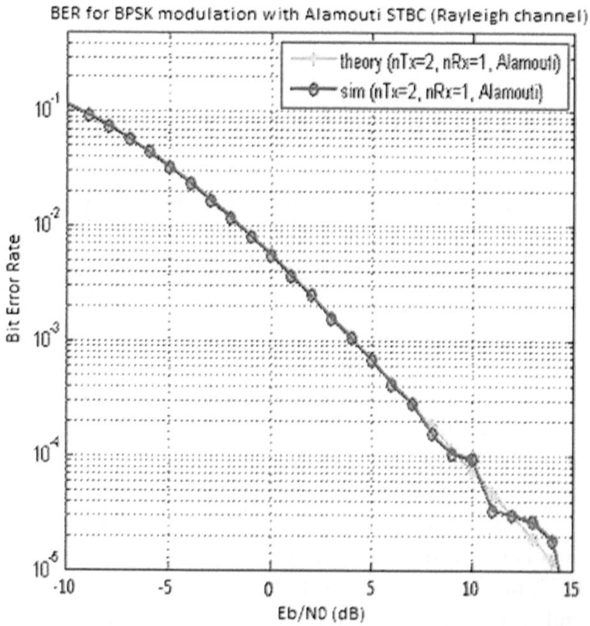

Fig. 2.4 BER of BPSK modulation with Alamouti STBC (Rayleigh channel)

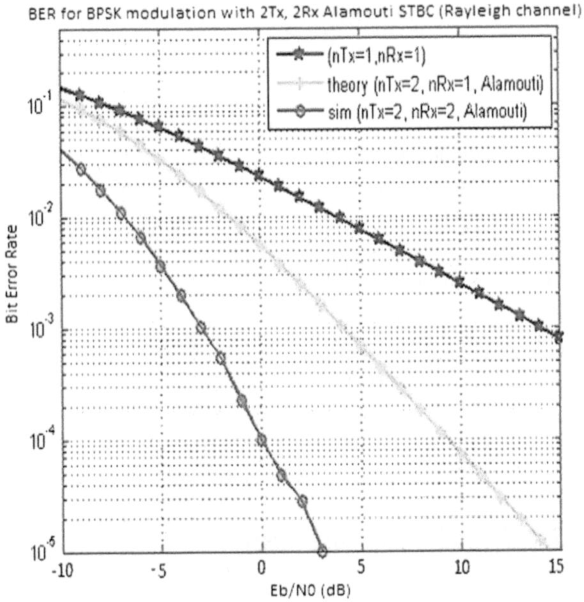

Fig. 2.5 BER of BPSK modulation with 2Tx and 2Rx Alamouti STBC (Rayleigh channel)

Table 2.2 BER performance at AWGN, Rayleigh, and MIMO channel

BER in dB	AWGN channel	Rayleigh channel	MIMO channel
−10	$10^{-0.5}$	$10^{-2.5}$	$10^{-0.9}$
5	10^{-1}	$10^{-0.5}$	$10^{-1.5}$
10	10^{-3}	$10^{-1.5}$	10^{-4}

2.3.3 Application of ANN for STBC Decoding

The ANN can be used for STBC decoding. It can also provide estimation of the channel which may help to mitigate some of the deficiencies of multi-user transmission. In ANN based systems, the multipath fading can be minimized by training the ANN and we can have a system with better bit error rates (BERs).

The ANN as multilayer perceptron (MLP) is trained using BP algorithm which updates the connecting weights. This adaptive updating of the MLP is continued till the performance goal is met. To train the ANNs, the setup we have used the received signal as the input signal and the transmitted signal is the reference signal or target. Tables 2.3 and 2.4 give the details of ANN setup.

Table 2.3 Details of ANN setup

ANN	MLP
Data set size	Training—10000 Testing—10000
Training type	TRAINGDX
Maximum number of epochs	1500
Variance in training data	50 %

Table 2.4 Performance of ANN

Networks	Data set size	Epochs	Time (s)	MSE
Net 1	10,000	1380	90	10^{-7}
Net 2	10,000	1456	120	10^{-7}
Net 3	10,000	950	80	10^{-7}

2.3.4 Result and Discussion

Here the modulated data is encoded in STBC encoder and sent through the two transmitters. The received data is decoded using ANN-based STBC decoding. The decoded data is demodulated and BER performance is evaluated. The ANN is trained with 10,000 b. The ANN is given a performance goal of around 10^{-3} which is attained after certain number of sessions, though the time taken is around 75 s. Figures 2.6 and 2.7 show the BER values of BPSK and QPSK signal using ANN for STBC decoding in Rayleigh fading channel. The values of the SNR varies from −10 to +20 dB.

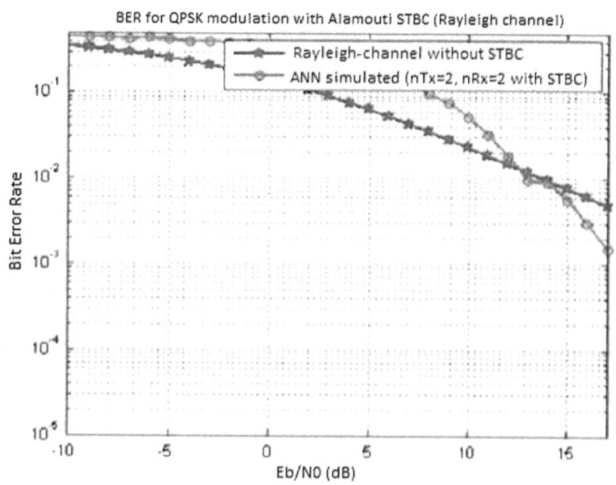

Fig. 2.6 BER plot of QPSK signals in MIMO set-up using STBC code in Rayleigh

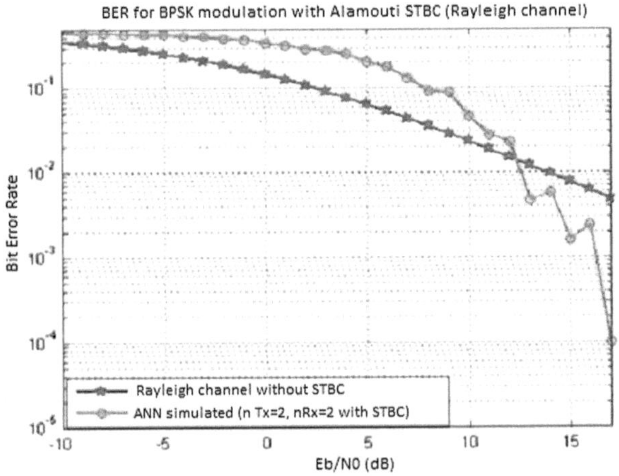

Fig. 2.7 BER plot of BPSK signals in MIMO set-up using STBC code in Rayleigh

2.4 Conclusion

STBC is used to achieve the maximum diversity order for the given number of transmit and receive antennas subject to the constraint of having a simple linear decoding algorithm. One of the viable means of better STBC decoding is the use of soft computing tools like the ANNs. In this work, we have analyzed and simulate the MIMO system with different modulation schemes and STBC in Rayleigh fading channel. We have used ANN for the decoding of STBC coded signal. The application of ANN in Rayleigh fading channel can improve the performance in a wireless communication system. The performance derived makes the approach a reliable means for study and analysis of the design of reception methods of MIMO system.

References

1. Foschini GJ, Gans MJ (1998) Limits of wireless communications in a fading environment when using multiple antennas. Wirel Pers Commun 6:311–335
2. Foschini GJ (1996) Layered space-time architecture for wireless communications in a fading environment when using multi-element antennas. Bell Labs Tech J 1:41–59
3. Telatar E (1999) Capacity of multi-antenna Gaussian channels. Eur Trans Telecommun 10(6):585–595
4. Mouly M, Pautet MB (1992) The GSM systems for mobile communications. Telecommunications Publishing, Olympia
5. Asif ZS (2004) Mobile receive diversity technology improves 3G systems capacity. In: Proceedings of IEEE Radio and Wireless Conference, pp 7803–8451
6. Hiwale AS, Ghatol AA (2007) Capacity and performance analysis of space-Time Block Codes in Rayleigh Fading Channels. Wseas Trans Commun 6(12):861–866
7. Sarma KK, Mitra A (2009) ANN based Rayleigh multipath fading channel estimation of a MIMO-OFDM system. In: Proceedings of 1st IEEE Asian Himalayas international conference on Internet (AH-ICI), pp 1–5

8. Popa S, Draghiciu N, Reiz R (2008) Fading types in wireless communications systems. J Electr Electr Eng 1(1):232–236
9. Tarokh V, Seshadri N, Calderbank AR (1998) Space-time codes for high data rate wireless communication performance criterion and code construction. IEEE Trans Inf Theory 44(2):744–765
10. Alamouti SM (1998) A simple transmitter diversity scheme for wireless communications. IEEE J Selected Areas Commun 16:1451–1458
11. Tarokh V, Jafarkhani H, Calderbank AR (1999) Space-time block codes from orthogonal designs. IEEE Trans Inf Theory 45(5):1456–1467

Chapter 3
Carrier Phase Detection of Faded Signals Using Digital Phase-Locked Loop

Basab Bijoy Purkayastha and Kandarpa Kumar Sarma

Abstract In this chapter, the design of a digital receiver for carrier phase tracking is presented. The receiver architecture includes a least square polynomial fitting (LSPF)-based digital phase-locked loop (DPLL). Bit error rate (BER) performance of the proposed system for dealing with Rayleigh and Rician fading for different numbers of paths with coded and uncoded channel is presented here. The performance of the DPLL for carrier phase tracking with signal using QPSK modulation transmitted through Rayleigh and Rician fading channels are compared with coded and uncoded conditions. Simulation results show that the proposed DPLL-based approach shows significant improvement using BCH coding both in Rayleigh and Rician fading channels. Several essential processes like noise and CCI cancelation, equalization, etc., that are integral to the traditional frameworks are made redundant by the proposed DPLL-based approach. The composite outcome of these separate processes is combined by the DPLL action making it a reliable and efficient mechanism leading to a compact design.

Keywords Bit error rate (BER) · Carrier phase detection · Digital phase-locked loop (DPLL) · Least square polynomial fitting (LSPF) · QPSK modulation · Nakagami-m model

B.B. Purkayastha (✉)
Department of Physics, IIT Guwahati, Guwahati, Assam, India
e-mail: basab.bijoy@gmail.com

K.K. Sarma
Department of Electronics and Communication Technology,
Gauhati University, Guwahati 781014, Assam, India
e-mail: kandarpaks@gmail.com

K.K. Sarma et al. (eds.), *Recent Trends in Intelligent and Emerging Systems*, Signals and Communication Technology, DOI 10.1007/978-81-322-2407-5_3

3.1 Introduction

The phase-locked loop (PLL) principles have already been employed during the last few decades for coherent detection. PLL can be described to be a receiver component that extracts the information of the frequency and the phase of the input signal precisely and generate a phase error signal. Based on this signal, it adjusts the frequency and phase of the local oscillator. PLLs generate stable frequencies, recover a signal from a noisy communication channel, or distribute clock timing pulses in digital logic designs such as microprocessors. Since a single integrated circuit (IC) can provide a complete PLL building block, the technique is widely used in modern electronic devices, with output frequencies from a fraction of a hertz up to many gigahertz.

It is well known that carrier recovery, which is necessary for coherent detection, suffers from the time-variant channel conditions in a multipath fading environment. Random frequency modulation is the main reason why traditional PLL-based structures for carrier recovery, do not operate satisfactorily. Therefore, we have proposed the modified structure of a digital phase-locked loop (DPLL) for carrier detection, dealing with Nakagami-m fading environment [1]. We have derived bit error rate (BER) performance of the proposed DPLL under varied fading conditions including Nakagami-m model using QPSK modulation and uncoded conditions. Some of the relevant literatures are [1–18]. It is found to be comparable to the performance of the existing systems of similar type [2–4].

The use of least square (LS) digital polynomial fitting filters (LSDPFF) to reduce random noise in time or wavelength variant analytical data has become widespread in the last few decades since Savitzky and Golay published the concept [5, 6]. The polynomial fitting method eliminates the phase noise in the continuous interval. In addition to applying these filters for increasing signal-to-noise ratio (SNR) with minimum signal distortion, they are also extremely useful in the numerical differentiation of data, producing results which are relatively insensitive to high-frequency noise. Here, we have utilized this LS digital polynomial fitting tool not only for increasing the carrier-to-noise ratio (CNR), but also as an integral part of the proposed DPLL to accomplish the necessary phase and frequency tracking of the incoming faded noisy carrier. Fading is an important factor in wireless medium contributing to signal degradation and error performance for all modulation schemes. For most practical channels, where signal propagation takes place in the atmosphere and near the ground, the free-space propagation model is inadequate to describe the channel and predict system performance. Depending on the environments of the communication system, many channel models have been proposed for the statistical description of the amplitude and phase of multipath fading signals. Out of these models, Rayleigh, Rician, and Nakagami Fading models are most widely used [7–11]. The proposed DPLL-based approach is designed to investigate its performance over wireless multipath under severe to less fading conditions coupled with additive noise. The importance of considering channel coding for error detection is apparent in the design and analysis of communication systems. Channel coding for error detection and correction helps the communication system designers to minimize the effects of a noisy multipath transmission channel [12, 13].

We present here the implementation and related results of BER performance analysis of the DPLL-based system for dealing with Rayleigh and Rician fading for different numbers paths using BCH(15,7)-based channel coding error detection and correction technique. Results are also compared with the performance of the DPLL-based system using uncoded transmission mechanism. Many process like noise and CCI cancelation, equalization, etc., which are essential for successful data recovery in wireless communication are combined by the DPLL action making it a reliable and efficient mechanism leading to compact design.

This chapter is organized into the following sections: Sect. 3.2 provides the background, motivation, and governing principles that have been considered while carrying out the work. A brief discussion has been made on QPSK Signal, BCH(n,k) coding, decoding, and Rayleigh and Rician fading channel modeling in the section with all relevant details. Section 3.3 of this paper briefly describes the already proposed DPLL structures and functionalities. The experimental details and the results derived constitute Sect. 3.4. The results obtained for different fading conditions and signal-to-noise ratios (SNRs) are also given. Section 3.5 concludes the description.

3.2 Background Considerations

In this section, we present briefly the related notions of QPSK, BCH(15,7) coded channels and Rayleigh and Rician fading. The following subsections provide certain theoretical aspects.

3.2.1 *QPSK Signal Modeling*

We have chosen QPSK signal modulation scheme in order to evaluate the performance of the proposed DPLL. QPSK is the most widely used phase modulation scheme and has applications that range from voice-band modems to high-speed satellite transmissions. The QPSK signals are defined as follows:

$$s_i(t) = A\cos(2\pi f_c t + \theta_i) \qquad 0 \le t \le T, \qquad i = 1, 2, 3, 4 \tag{3.1}$$

where $\theta_i = (2i - 1)\pi/4$. The four available phases are therefore $\pi/4$, $3\pi/4$, $5\pi/4$, and $7\pi/4$.

3.2.2 BCH(15,7) Coding and Decoding

BCH codes are a large class of cyclic codes that include both binary and nonbinary codes. Binary (n, k), with any positive integer $m \geq 3$, BCH codes can be constructed with the following parameters

$$n = 2^m - 1, \qquad n - k \leq mt, \qquad d_{\min} \geq 2t + 1 = \delta \tag{3.2}$$

where t is the error correcting capability and δ is called the code design distance. It means that a BCH code with specified parameters given in Eq. 3.2, guarantees to correct t or less number of errors in the received n block bits. We have chosen BCH(15,7) schemes, which has a code length of 15 and message length of 7. It can be used to correct a maximum of 2 errors out of 15 code bits. There are many algorithms for decoding BCH codes, which are already in circulation in open literature [12, 13].

3.2.3 Modeling Rayleigh and Rician Fading Channel

In wireless communications, signal fading is caused by multipath effect. Multipath effect means that a signal transmitted from a transmitter may have multiple copies traversing different paths to reach a receiver. Thus, at the receiver, the received signal should be the sum of all these multipath signals. Because the paths traversed by these signals are different, some are longer and some are shorter. The one at the direction of light of signal (LOS) should be the shortest. These signals interact with each other. If signals are in phase, they would intensify the resultant signal; otherwise, the resultant signal is weakened due to out of phase. This phenomenon is called channel fading.

3.2.3.1 Rayleigh Fading

In Rayleigh channel, the receiver, instead of receiving the signal over one line-of-sight path, receives a number of reflected and scattered waves with varying path lengths, the phases are random, and as a result of which, the instantaneous received power becomes random.

With no direct path or LOS component at the receiver, the Rayleigh channel samples $h(t)$ have been generated from the following expression

$$h(t) = \sum_{i=1}^{N} a_i \cos(\omega_c t + \phi_i) \tag{3.3}$$

where N is the number of paths. The phase θ_i depends on the varying path lengths, and are uniformly distributed over $[0, 2\pi]$. The parameter which affects data transmission the most in the context of small scale fading is the Doppler frequency due to relative motion between the transmitter and the receiver. The Doppler frequency can be expressed as

$$\omega_{di} = \frac{\omega_c v}{c} \cos \psi_i \tag{3.4}$$

where v is the velocity of the mobile, c is the speed of light (3×10^8) m/s, and the ψ_i are uniformly distributed over $[0, 2\pi]$. To include the effects of motion induced Doppler frequency and phase shifts the Eq. 3.3 must be modified as

$$s(t) = \sum_{i=1}^{N} a_i \cos(\omega_c t + \omega_{di} t + \phi_i) \tag{3.5}$$

Let, $x(t)$ be the transmitted QPSK modulated signal sample and $a(t)$ be the AWGN, then the received signal $r(t)$ can be given by

$$r(t) = h(t) * x(t) + a(t) \tag{3.6}$$

To evaluate the first-order statistics of the received signal, we have to decompose signal to in phase and quadrature component and we can write

$$r(t) = I(t) \cos \omega_c t - Q(t) \sin \omega_c t \tag{3.7}$$

The envelope of the received signal can be given by

$$R = \sqrt{[I(t)]^2 + [Q(t)]^2} \tag{3.8}$$

The probability distribution function Rayleigh distribution is given by:

$$f(r) = \frac{r}{\sigma^2} \exp\left(-\frac{r^2}{2\sigma^2}\right) \qquad r \geq 0 \tag{3.9}$$

3.2.3.2 Rician Fading

Rician fading condition can be describe as the presence of a direct path or line-of-sight component, in addition to the multipath components. In such a scenario, the transmitted signal given in Eq. 3.5 can be rewritten as

$$s(t) = \sum_{i=1}^{N-1} a_i \cos(\omega_c t + \omega_{di} t + \phi_i) + k_d \cos(\omega_c t + \omega_d t) \tag{3.10}$$

where the constant k_d is the strength of the LOS component, ω_d is the Doppler shift along the LOS path, and ω_{di} are the Doppler shifts along the indirect paths given by Eq. 3.4. The probability distribution function Rician distribution is given by:

$$f(r) = \frac{r}{\sigma^2} \exp\left(-\frac{r^2 + {k_d}^2}{2\sigma^2}\right) I_0\left(\frac{rk_d}{\sigma^2}\right) \qquad r \geq 0 \tag{3.11}$$

where $I_0(.)$ is the zero-order-modified Bessel function of the first kind. The cumulative distribution of the Rician random variable is given as

$$F(r) = 1 - Q\left(\frac{K_d}{\sigma}, \frac{r}{\sigma}\right) \qquad r \geq 0 \tag{3.12}$$

where $Q(,)$ is the Marcumas Q function. The Rician distribution is often described in terms of the Rician factor K, defined as the ratio between the deterministic signal power (from the direct path) and the diffuse signal power (from the indirect paths). K is usually expressed in decibels as

$$k(dB) = 10 \log_{10}\left(\frac{{k_d}^2}{2\sigma^2}\right) \tag{3.13}$$

In Eq. 3.13, if k_d goes to zero (or if $\frac{{k_d}^2}{2\sigma^2} \ll \frac{r^2}{2\sigma^2}$), the direct path is eliminated and the envelope distribution becomes Rayleigh, with $K(dB) = \infty$.

3.2.3.3 Statistical Characterization of Fading Channels Using Nakagami-*m* Distribution

It is possible to describe both Rayleigh and Rician fading with the help of a single model using the Nakagami distribution. Nakagami fading assumes that the transmitted signal that has passed through the channel will fade according to Nakagami distribution [14]. The probability distribution function of the received signal envelope in terms of Nakagami-m distribution is given by:

$$p(r) = \frac{2m^m r^{2m-1}}{\Gamma(m)\Omega^m} \exp\left(-\frac{mr^2}{\Omega_p}\right), \qquad r \geq 0 \tag{3.14}$$

where r is Nakagami envelope, $\Gamma(m)$ is the Gamma function, and m is the shape factor (with the constraint that $m \geq \frac{1}{2}$) given by

$$m = \frac{E(r^2)}{var(r^2)} = \frac{E^2\left(r^2\right)}{E\left([r^2 - E(r^2)]^2\right)} \tag{3.15}$$

The parameter Ω is the instantaneous power that controls the spread of the distribution and is given by

$$\Omega = E\left(r^2\right) \tag{3.16}$$

The significance of this adaptive m parameter are as follows:

1. If the envelope is Nakagami distributed, the corresponding power is Gamma distributed.
2. In the special case $m = 1$, Rayleigh fading is recovered, with an exponentially distributed instantaneous power.
3. For $m > 1$, the fluctuations of the signal strength are reduced as compared to Rayleigh fading.
4. For $m = 0.5$, it becomes one-sided Gaussian distribution.
5. For $m = \infty$, the distribution becomes impulse, i.e., no fading. The sum of multiple independent and identically distribute Rayleigh-fading signals has Nakagami distributed signal amplitude.

To statistically characterize fading channel, the level-crossing rates (LCRs) reflect the scattering environment, and thus are called the second-order statistics of a fading channel. The LCR is defined as the number of times per second that the envelope of fading channel crosses a specified level Y_0 in a positive-going direction, then average number of positive zero crossings can be given by the expression

$$N_{Y_0} = \int_0^\infty (Y^/ p(Y_0, Y^/)dY^/ \tag{3.17}$$

and it can be rewritten as

$$N_{Y_0} = \frac{2\sigma Y^/}{\sqrt{2\pi}\Gamma(m(r))}\left(\frac{m}{\Omega_p}\right)^m {Y_0}^{2m-1}\exp\left(\frac{m{Y_0}^2}{\Omega_p}\right) \tag{3.18}$$

3.3 Proposed Digital Phase-Locked Loop

The modified structure of the DPLL has three major components, namely, least square polynomial fitting (LSPF) block, roots approximator (RA), and numerically controlled oscillator (NCO). The block diagram of the DPLL is shown in the Fig. 3.1 with dashed boundary. The system performs using uniform sampling with moderate sampling frequency. The proposed DPLL performs in piecewise manner. It accepts signal samples for one symbol period at a time, and executes subsequent processing. Here, we include a brief description of the functionalities of the major components of the DPLL.

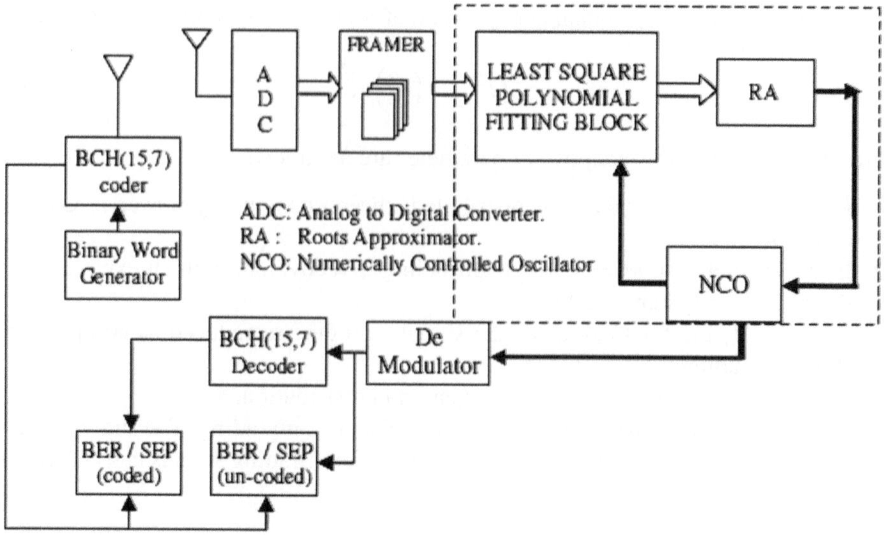

Fig. 3.1 DPLL-based carrier and symbol recovery system for Rayleigh and Rician channels

We replace the traditional phase frequency detector (PFD) of a DPLL with LSPF block because, it can take care of functionalities of two major components of a traditional DPLL, namely signal conditioning and phase frequency detection. It has two inputs, the frequency count of local reference signal and the input signal. We have found that the sixth-order polynomial fitting can generate an acceptable estimate of the QPSK modulated signal under effect of additive noise and multipath path fading channel [1, 15].

If $(t_1, y_1), (t_2, y_2), \ldots, (t_n, y_n)$, represents the faded signal samples then by applying LSPF, faded signal samples can be fitted to best-fit polynomial of degree six, provided $n \geq 7$ so that the sum of squared residuals S is minimized.

$$\hat{y} = a_0 + a_1 t + a_2 t^2 + a_3 t^3 + a_4 t^4 + a_5 t^5 + a_6 t^6 \tag{3.19}$$

$$S = \sum_{i=1}^{n} [y_i - \hat{y}]^2 = \sum_{i=1}^{n} [y_i - (a_0 + a_1 t + \cdots + a_6 t^6]^2. \tag{3.20}$$

By obtaining the partial derivatives of S with respect to $a_0, a_1, a_2, \ldots, a_6$ and equating these derivatives to zero, the following matrix equation is defined:

$$\begin{pmatrix} n & \sum_{i=1}^{n} t_i & \sum_{i=1}^{n} t_i{}^2 & \cdots & \sum_{i=1}^{n} t_i{}^6 \\ \sum_{i=1}^{n} t_i & \sum_{i=1}^{n} t_i{}^2 & \sum_{i=1}^{n} t_i{}^3 & \cdots & \sum_{i=1}^{n} t_i{}^7 \\ \sum_{i=1}^{n} t_i{}^2 & \sum_{i=1}^{n} t_i{}^3 & \sum_{i=1}^{n} t_i{}^4 & \cdots & \sum_{i=1}^{n} x_i{}^8 \\ \sum_{i=1}^{n} t_i{}^3 & \sum_{i=1}^{n} t_i{}^4 & \sum_{i=1}^{n} t_i{}^5 & \cdots & \sum_{i=1}^{n} x_i{}^9 \\ \sum_{i=1}^{n} t_i{}^4 & \sum_{i=1}^{n} t_i{}^5 & \sum_{i=1}^{n} t_i{}^6 & \cdots & \sum_{i=1}^{n} x_i{}^{10} \\ \sum_{i=1}^{n} t_i{}^5 & \sum_{i=1}^{n} t_i{}^6 & \sum_{i=1}^{n} t_i{}^7 & \cdots & \sum_{i=1}^{n} x_i{}^{11} \\ \sum_{i=1}^{n} t_i{}^6 & \sum_{i=1}^{n} t_i{}^7 & \sum_{i=1}^{n} t_i{}^8 & \cdots & \sum_{i=1}^{n} t_i{}^{12} \end{pmatrix} \begin{pmatrix} a_0 \\ a_1 \\ a_2 \\ a_3 \\ a_4 \\ a_5 \\ a_6 \end{pmatrix} = \begin{pmatrix} \sum_{i=1}^{n} y_i \\ \sum_{i=1}^{n} t_i y_i \\ \sum_{i=1}^{n} t_i{}^2 y_i \\ \sum_{i=1}^{n} t_i{}^3 y_i \\ \sum_{i=1}^{n} t_i{}^4 y_i \\ \sum_{i=1}^{n} t_i{}^5 y_i \\ \sum_{i=1}^{n} t_i{}^6 y_i \end{pmatrix} \tag{3.21}$$

From this equation the set of coefficients $a_0, a_1, a_2, \ldots, a_6$, which are the unique solution of this system are obtained.

The coefficients of best-fit polynomial function of noisy faded signal are measures of the phase and frequency associated with that signal. LSPF does three jobs. *First*, it equates the coefficients of best-fit polynomial function to the incoming faded signal. *Second*, with the use of the coefficients of polynomial function it generates the best-fit signal samples free from ripples. *Third*, LSPF feeds the value coefficients of best-fit polynomial function and the frequency count of local reference signal to RA for further processing. The RA is the next major component in the proposed DPLL. It takes the value of the coefficients of best-fit polynomial function and computes the roots of that polynomial function. Roots of the polynomial function carry the phase and frequency information of the fitted signal. Given any degree polynomial, the roots can be found by finding the eigenvalues of the following matrix

$$\begin{bmatrix} -a_1/a_0 & -a_2/a_0 & -a_3/a_0 & \cdots & -a_6/a_0 \\ 1 & 0 & 0 & \cdots & 0 \\ 0 & 1 & 0 & \cdots & 0 \\ \vdots & \vdots & \vdots & \ddots & \vdots \\ 0 & 0 & 0 & \cdots & 0 \end{bmatrix} \tag{3.22}$$

And then corresponding roots can be computed by the following expression

$$r_i = \frac{1}{\lambda_i} \tag{3.23}$$

After evaluating the roots, the RA performs three additional jobs. It arranges the roots in ascending order and calculates the time period of fitted signal from difference of any two alternate sorted roots values. From this time period, RA calculates the frequency associated with fitted signal. Among the sorted roots, RA computes the first positive to negative zero crossing root, which signifies the phase associated with fitted signal samples. RA feeds two information namely frequency count, phase count to NCO for further processing. The NCO takes phase and frequency information from RA as input and adjust its local reference signal's phase and frequency and outputs a new reference signal. As the DPLL is proposed for QPSK modulation scheme, it is expected that the frequency of the signal will not vary much, only the phase of the carrier associated with each symbol will vary. So we propose a new type of NCO having two outputs. The first output will be frequency corrected zero-phased sinusoidal signal, and the second output will be frequency as well as phase corrected signal which will be applied as demodulator input.

3.4 Experimental Considerations, Results, and Discussion

An experimental model is created integrating each block as described in Sect. 3.3, which represents a complete communication scenario. QPSK signal is generated by modulating a sufficient numbers of random binary bits coded with BCH(15,7) schemes with a carrier and transmitted over Rayleigh and Rician channel separately with varying numbers of path.The system has been simulated for counting bit errors occurred during reception and demodulation of 75×10^6 transmitted BCH(15,7)-coded binary bits with carrier frequency 900 MHz. We have generated different sets of Rayleigh and Rician channel with number of paths varied from 1 to 10. In case of Rician channel, one LOS is considered in addition to the indirect paths. Then, each of these sets is multiplied with signal samples to result-faded signal sets of different fading figure. Now to each of the faded signal set, AWGN is added with SNR value ranging from 0 to 30 dB to produce further multiple sets of faded noisy signal sets, each representing combination of different fading conditions and different SNRs of received signal. The received signal samples are allowed to pass through the DPLL. Output of DPLL is converted back to binary bits after necessary demodulation. Bit Errors occurred are counted. Demodulated binary bits are then passed through BCH(15,7) decoder for error detection and correction. From this decoded form, binary bits are again compared to the modulating binary bits, to count the bit errors after error detection and correction.

Table 3.1 summarizes the variation of Nakagami fading parameter (m) and corresponding mean instantaneous power (Ω) at different fading environment expressed in terms of Rayleigh and Rician distribution with different number of multipath components. In case of both Rayleigh and Rician channel, it can be seen that the value of (m) increases with the increase in SNR, i.e., with the reduction in the noise power (AWGN) the environment becomes more favorable for communication. The tabulated data also indicates the same, e.g., for 10-path Rayleigh channel at 0 dB, m takes a value of 1.011 and it increases to 1.251 at 30 dB, same is the case for all other combination of channels including Rician too. Again form the Table 3.1, it can also be inferred that with an increase in the number of multipath components the communication environment degrades. This is due to the fact that, all those multipath

Table 3.1 Nakagami parameter (m) and mean power (Ω) for Rayleigh and Rician channel at various SNRs

Case		Nakagami (m) and mean power (Ω) at various SNRs						
		0 dB	5 dB	10 dB	15 dB	20 dB	25 dB	30 dB
Rayleigh	m	1.202	1.423	1.573	1.621	1.632	1.656	1.651
1 Path	Ω	1.30	1.00	0.91	0.88	0.87	0.87	0.87
Rayleigh	m	1.201	1.363	1.517	1.554	1.563	1.571	1.574
5 Path	Ω	2.94	2.26	2.04	1.99	1.96	1.95	1.95
Rayleigh	m	1.011	1.053	1.122	1.154	1.193	1.237	1.251
10 Path	Ω	4.03	3.01	2.68	2.59	2.56	2.55	2.55
Racian	m	1.79	2.95	4.75	5.98	6.79	7.82	8.25
1 Path	Ω	3.93	3.12	2.85	2.77	2.76	2.75	2.74
Racian	m	1.748	2.235	4.034	5.121	6.072	7.011	7.413
5 Path	Ω	3.96	3.15	2.89	2.82	2.79	2.79	2.78
Racian	m	1.661	2.013	3.551	4.632	5.215	5.337	5.402
10 Path	Ω	4.10	3.25	2.98	2.90	2.87	2.87	2.86

components arrived at the receiver with different time delays and path gains hence making the resultant received signal degraded incorporating phase and magnitude noise. The results obtained in terms Nakagami fading parameter (m) also supports this facts. In case of Rician channel, the value of (m) is much higher than its Rayleigh counterpart. This is due to the contribution from the LOS components among other multipath components (Fig. 3.2).

The PDF of received signal's envelope for Rayleigh channel with different numbers of multipath component (1, 5, and 10) and at different SNRs are represented by the left-hand side plots of Fig. 3.3a, c, e. The corresponding plots of LCR are represented by the right-hand side plots of same Fig. 3.3b, d, f. It is clear from the plots that with the increase of m, which means improvement in the propagation environment, the LCR gradually decreases. Figure 3.4 represents the same for Rician channel case. BER performance of the DPLL is represented by a BER versus SNR.

Figure 3.5a–f for plots shown in Rayleigh and Rician channel with different number of multipath components (1, 5, and 10) and various combinations of SNR both under coded and uncoded channel situations. From the plots, we can conclude that under BCH(15,7)-coded channel case system performance improves significantly as compared to uncoded channel case.

In recent years, the results of BER analysis for QPSK signal in Rayleigh and Rician Fading Channel have been reported by a few authors [16–18]. Although the approach they have proposed for the recovery of the signal is not DPLL based, in comparison to their works our DPLL-based approach have certainly shown superior performance in terms of BER performance achieved for all kind channel conditions that we have experimented. We have compared the results with work reported in

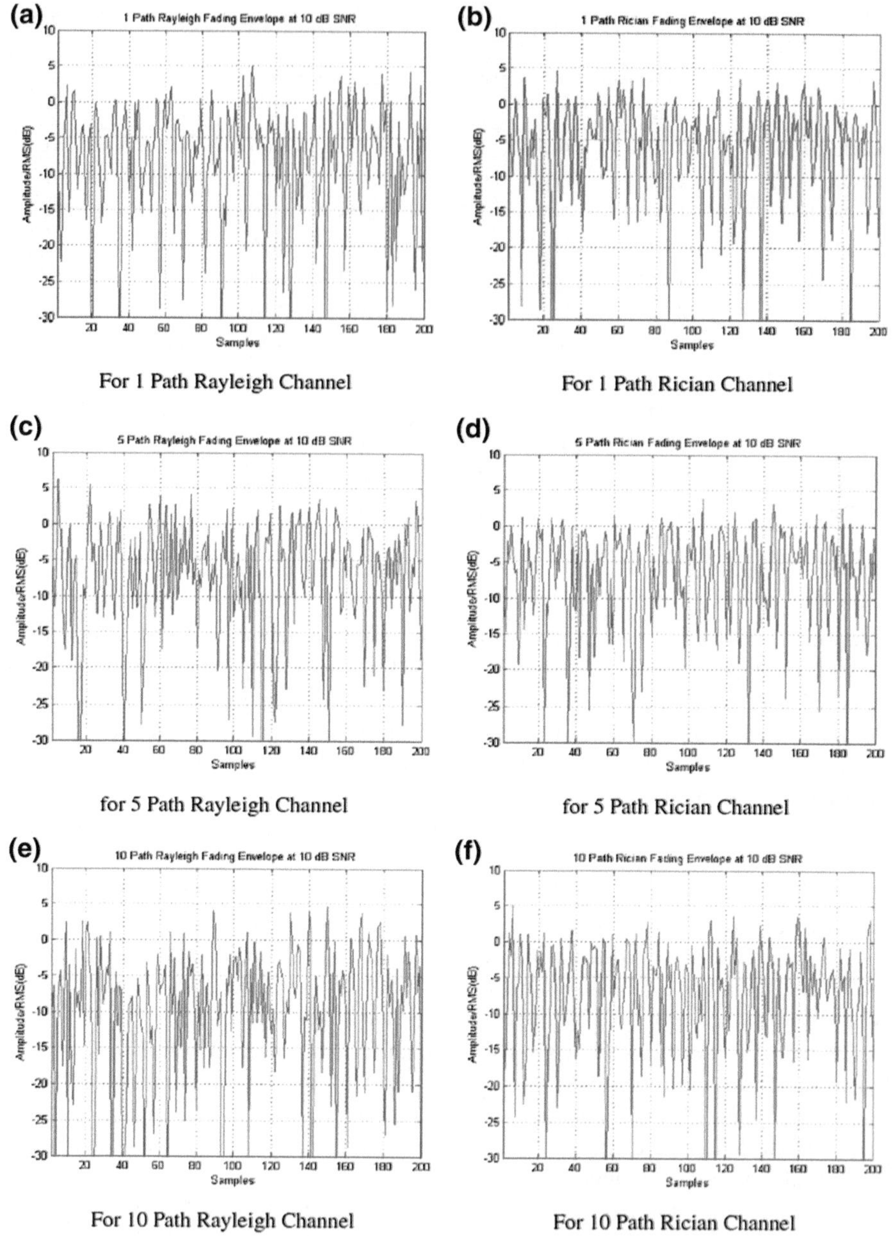

Fig. 3.2 Rayleigh and Rician channel envelope at 10 dB SNR

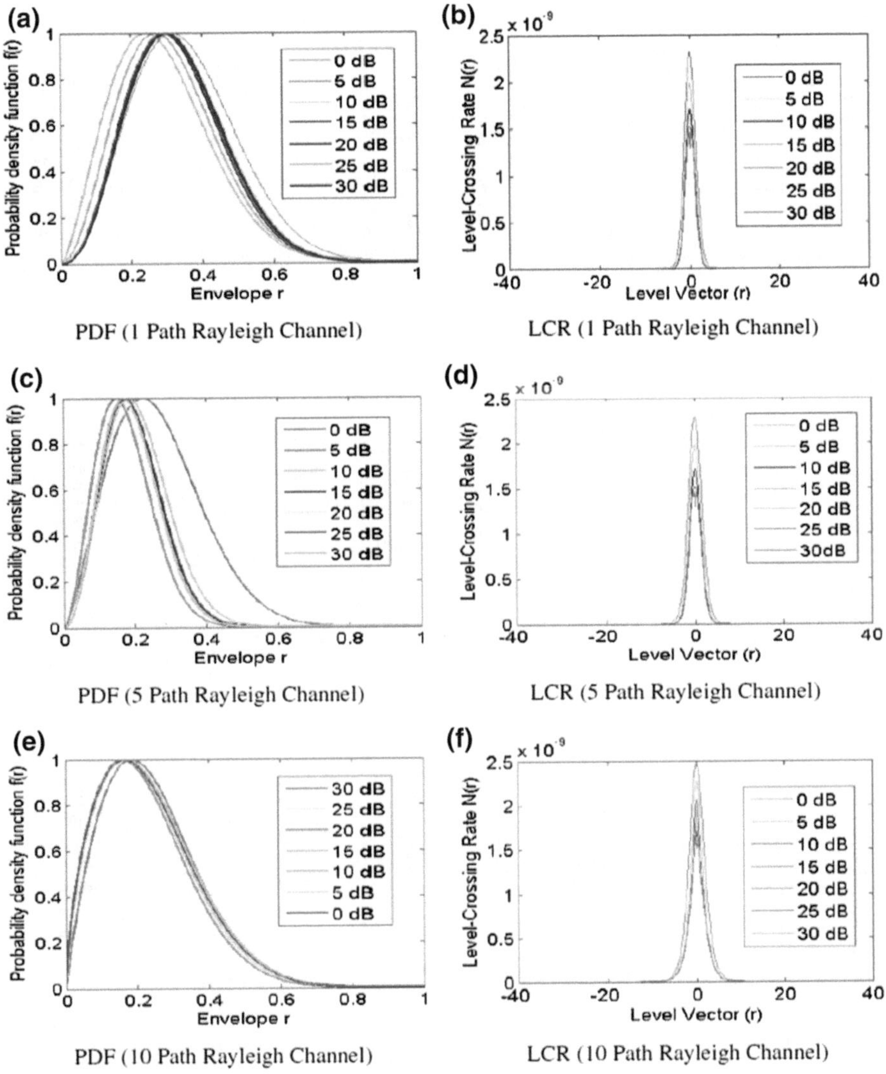

PDF (1 Path Rayleigh Channel) LCR (1 Path Rayleigh Channel)

PDF (5 Path Rayleigh Channel) LCR (5 Path Rayleigh Channel)

PDF (10 Path Rayleigh Channel) LCR (10 Path Rayleigh Channel)

Fig. 3.3 PDF and LCR for Rayleigh channel at different SNR

Ref. [17] for Rayleigh channel case and Ref. [16] for Rician channel case. These comparisons are represented in the plots of Fig. 3.4a–f.

Tables 3.2 and 3.3 summarizes the coding gain achieved upon the application of BCH(15,7)-based coding and decoding technique for various SNRs and different numbers of multipath components over Rayleigh and Rician channel, respectively. Experimental results indicates that in case of both Rayleigh and Rician channel, at lower SNRs the coding gain increases sharply with an increase in SNR. At

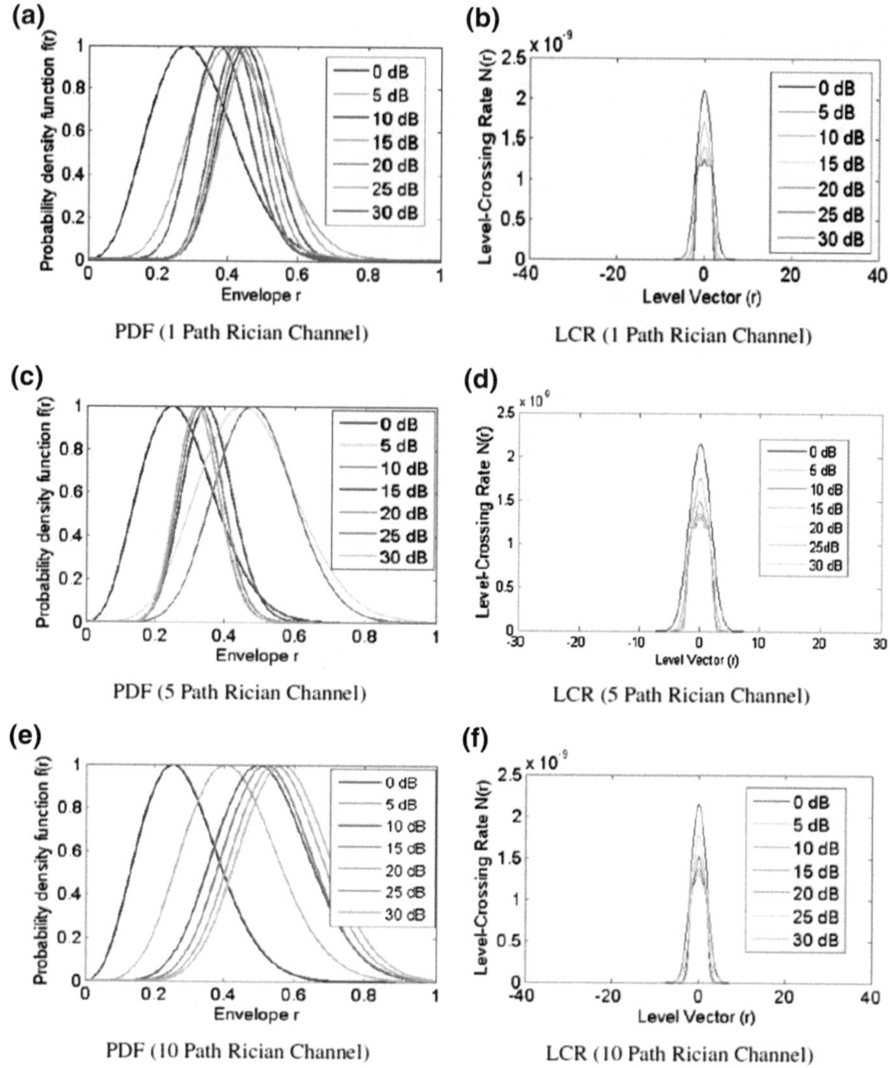

Fig. 3.4 PDF and LCR for Rician channel at different SNR

higher SNRs, the improvement is marginal. If we compare the respective coding gain achieved over Rayleigh and Rician channel with identical multipath component, it is found that at Rician channel the gain is much higher than its Rayleigh counterpart at lower SNRs (0 dB, 5 dB). At higher SNRs, the improvement is again marginal. The results obtained from the experiments also indicates that with a decrease in the number of multipath component at the receiver there is a marginal improvement in the coding gain in both Rayleigh and Rician channel. So, in the last row of both the Tables 3.2 and 3.3 we have provided the average coding gain achieved upon

Table 3.2 BCH(15,7) coding gain archived in percentage (%) for Rayleigh channel at various SNRs

Rayleigh channel	BCH(15,7) Coding gain in percentage (%) at various SNRs						
Parameter	0 dB	5 dB	10 dB	15 dB	20 dB	25 dB	30 dB
Rayleigh 1 Path	26.18	53.10	78.48	91.09	94.34	95.47	97.13
Rayleigh 5 Path	25.30	53.07	71.22	85.69	91.89	93.88	95.13
Rayleigh 10 Path	21.04	48.50	69.41	82.62	89.34	92.52	92.78
Average gain	24.17	51.56	73.04	86.47	91.86	93.96	95.01

Table 3.3 BCH(15,7) coding gain archived in percentage (%) for Rician channel at various SNRs

Racian channel	BCH(15,7) Coding Gain in Percentage (%) at various SNRs						
Parameter	0 dB	5 dB	10 dB	15 dB	20 dB	25 dB	30 dB
Racian 1 Path	60.90	72.50	78.10	87.47	93.63	97.54	100.00
Racian 5 Path	62.30	69.03	78.68	85.95	93.02	96.62	98.08
Racian 10 Path	61.65	72.80	77.43	84.34	92.35	94.22	95.06
Average gain	61.62	71.44	78.07	85.92	93.00	96.13	97.71

application of BCH(15,7)-based coding and decoding technique at various SNRs ranging from 0 to 30 dB. The advantages of the proposed DPLL based approach in combination with BCH(15,7) coding is thus obvious.

We have recorded the phase error response of the system during reception of sufficient numbers of known QPSK modulated symbols corrupted under various fading conditions and channeling effects. Due to randomness of the channel that we have modeled, we have repeated the simulation for atleast 100 times and data are averaged accordingly. The average phase error occurred for various SNRs and different numbers of multipath components over Rayleigh and Rician channel, respectively, is tabulated in Table 3.4. From Table 3.4, it can be seen that in case of both Rayleigh and Rician channel, phase error decreases with an increase in SNR. Again in case of both the channels, with an increase in number of path variables, phase error increases. This can be justified by the fact that with the increase in the number of multipath component, co-channel interference becomes more active, incorporating phase noise to received signal. Further, we have also found that in terms of phase error performance, Rician channel is superior to that of the Rayleigh counterpart.

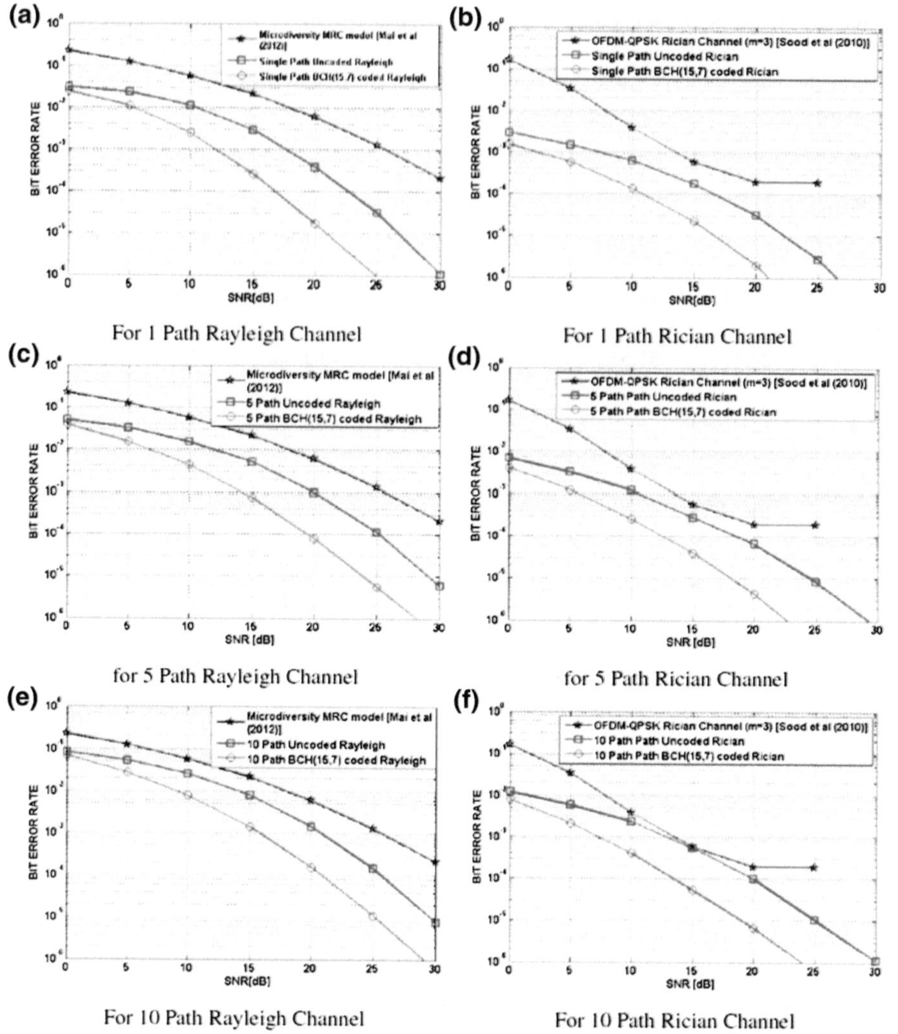

Fig. 3.5 BER versus SNR plot for Rayleigh and Rician channel at different SNR

Obvious reasons behind is that due to the presence of LOS component signals fidelity remains intact despite co-channel interference and channel conditions. The average results are shown in the Fig. 3.6.

Table 3.4 Average phase error detected from known transmitted symbol

Channel	Average phase error detected for known transmitted symbol						
Parameter	0 dB	5 dB	10 dB	15 dB	20 dB	25 dB	30 dB
Rayleigh 10 Path	0.6882	0.5042	0.3505	0.2501	0.1917	0.1422	0.1151
Rayleigh 5 Path	0.6240	0.4447	0.2989	0.19568	0.14010	0.0900	0.0706
Rayleigh 1 Path	0.5332	0.3666	0.2357	0.1509	0.0968	0.0515	0.0347
Racian 10 Path	0.3426	0.2573	0.1778	0.1123	0.0615	0.0335	0.0208
Racian 5 Path	0.2951	0.2142	0.1433	0.0857	0.0427	0.0198	0.0157
Racian 1 Path	0.2484	0.1760	0.1144	0.0639	0.0335	0.0121	0.0101

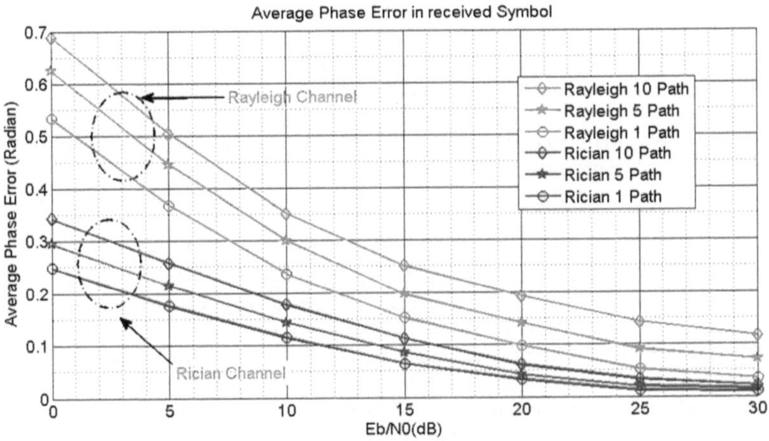

Fig. 3.6 Average phase error detected for known transmitted symbol

3.5 Conclusion

In this chapter, we have discussed about the implementation of a DPLL for carrier detection under different fading conditions using Rayleigh and Rician channel model. The performance of the system have been evaluated by varying numbers of multipath components for both type of channels. We have also statistically characterized the received signal in terms of PDF and LCR. Performance comparison have also been made for BCH(15,7) coded and uncoded QPSK modulated signal. It has been found that the systems performance improves significantly upon application of BCH channel coding. We have also seen that the proposed DPLL-based approach emerges to be better in terms of performance and compactness in design compared to other traditional approaches. Because, many process like noise and CCI cancelation, equalization, etc., which are essential for successful wireless communication are combined by the DPLL action making it a reliable and efficient mechanism leading to compact design.

References

1. Purkayastha BB, Sarma KK (2012) Digital phase locked loop based system for Nakagami fading channel model. Int J Comput Appl 42(9):1
2. Sood N, Sharma AK, Uddin M (2010) BER performance of OFDM-BPSK and -QPSK over generalized gamma fading channel. Int J Comput Appl 3(6):13–16
3. Smadi MA (2009) Performance analysis of QPSK system with Nakagami fading using efficient numerical approach. In: Proceedings of the IEEE 9th Malaysia international conference on communications, Kuala Lumpur, Malaysia, pp 447–450
4. Cheng J, Beaulieu NC, Zhang X (2005) Precise BER analysis of dual-channel reception of QPSK in Nakagami fading and cochannel interference. Proc IEEE Commun Lett 9(4):316–318
5. Savitzky A, Golay MJE (1972) Smoothing and differentiation of data by simplified least squares procedures. Anal Chem 44(X):1906–1909
6. Leach RA, Carter CA, Harris JM (1984) A least-squares polynomial filters for initial point and slope estimation. Anal Chem 56(13):2304–2307
7. Chavan MS, Chile RH, Sawant SR (2011) Multipath fading channel modeling and performance comparison of wireless channel models. Int J Electron Commun Eng 4(2):189–203
8. Weihua Z, Huang WV (1997) Phase precoding for frequency-selective Rayleigh and Rician slowly fading channels. IEEE Trans Veh Technol 46(1):129–142
9. Sudhir AB, Rao KVS (2011) Evaluation of BER for AWGN, Rayleigh and Rician fading channels under various modulation schemes. Int J Comput Appl 9(26):23–28
10. Chengshan X, Zheng YR, Beaulieu NC (2003) Statistical simulation models for Rayleigh and Rician fading. Proc IEEE Int Conf Commun 5:3524–3529
11. Luo JX, Zeidler JR (2000) A statistical simulation model for correlated Nakagami fading channel. In: Proceedings of the international conference on communications technology, vol 2. Beijing, China, pp 1680–1684
12. Hanfeng C, Gulliver TA, Wei L, Hao Z (2006) Performance of ultra-wideband communication systems using DS-SS PPM with BCH coding over a fading channel. In: Proceedings of IEEE military communications conference (MILCOM 2006), pp 1–5
13. Vardy A, Beery Y (1993) Maximum-likelihood soft decision decoding of Bch codes. In: Proceedings of IEEE international symposium on information theory, pp 17–22
14. Nakagami M (1960) The m-distribution, a general formula of intensity of rapid fading. In: Statistical methods in radio wave propagation. In: Hoffman WG (ed), Oxford Edn. Permagon Press, UK, pp 3–36
15. Purkayastha BB, Sarma KK (2012) A digital phase locked loop for Nakagami fading channels using QPSK modulation schemes. In: Proceedings of 2nd IEEE national conference on computational intelligence and signal processing. Guwahati, India, pp 141–146
16. Sood N, Chile, Sharma AK (2010) BER performance of OFDM-BPSK and QPSK over Nakagami-m fading channels. In: Proceedings of IEEE 2nd international advance computing conference, pp 88–90

17. Mai DTT, Cong LS, Tuan NQ, Nguyen D-T-T (2012) BER of QPSK using MRC reception in a composite fading environment. In: Proceedings of IEEE international symposium on communications and information technologies (ISCIT), pp 486–491
18. Zhang S, Kam PY, Ho PKM (2004) Performance of BPSK and QPSK over the nonselective Rayleigh fading channels with maximal ratio combining and multiple asynchronous cochannel interferers. In: Proceedings of IEEE international conference on communications, circuits and systems (ISCIT), vol 1, pp 126–130

Chapter 4
Adaptive MRC for Stochastic Wireless Channels

Atlanta Choudhury and Kandarpa Kumar Sarma

Abstract This work is related to design certain adaptive equalization and error correction coding and aids to maximal-ratio combining (MRC) in faded wireless channels. The performances derived are analyzed using semianalytic (SA) and Monte Carlo (MC) approaches in order to achieve improvement in bit error rates (BERs) of demodulated signals in wireless channels that have both Gaussian and multipath fading characteristics. Modulation techniques used in this work are bipolar phase shift keying (BPSK), quadrature phase shift keying (QPSK), and differential phase shift keying (DPSK). The work though considers the use of least mean square (LMS), adaptive filter blocks as part of a MRC setup and is tested under SNR variation between −10 and 10 dB. The results generated justify the use of the adaptive equalizer block as an aid to the MRC setup. The validity of the results is further confirmed by comparing to those obtained via SA approach and MC simulations.

Keywords Maximal-ratio combining (MRC) · Bit error rate (BER) · Least mean square (LMS) adaptive filter · Faded wireless channels · Bipolar phase shift keying (BPSK) · Quadrature phase shift keying (QPSK) · Differential phase shift keying (DPSK)

4.1 Introduction

Due to the rapid advancement of technology, wireless and mobile communications have seen unpredicted growth. As a result, bandwidth has become scarce and quality of service (QOS) has become a major factor in deciding the link reliability. The major

A. Choudhury (✉)
Department of Electronics and Telecommunication Engineering,
GIMT, Guwahati, Assam, India
e-mail: atlanta.ayush.choudhury@gmail.com

K.K. Sarma
Department of Electronics and Communication Technology,
Gauhati University, Guwahati 781014, Assam, India
e-mail: kandarpaks@gmail.com

K.K. Sarma et al. (eds.), *Recent Trends in Intelligent and Emerging Systems*,
Signals and Communication Technology, DOI 10.1007/978-81-322-2407-5_4

challenges faced are due to the fluctuations and fading observed in the propagation medium. In case of fading where there is an absence of line of sight (LOS), component Rayleigh fading comes into the picture. On the other hand, with a LOS segment, Rician fading is observed. Diversity combining is one of the most widely employed techniques in wireless communications receivers for mitigating the effects of multipath fading. It contributes significantly toward improving the overall system performance. The most popular diversity techniques are equal-gain combining (EGC), maximal-ratio combining (MRC), selection combining (SC), and a combination of MRC and SC, called generalized selection combining (GSC) [1, 2]. This work is related to the design of an MRC scheme assisted by an adaptive equalization block so as to investigate the performance of the MRC adaptive equalization combination in Gaussian and multipath slow fading channels. The work uses bipolar phase shift keying (BPSK), quadrature phase shift keying (QPSK), and differential phase shift keying (DPSK) modulation in Gaussian, Rayleigh, and Rician fading channels with signal-to-noise ratio (SNR) variation in the range -10 to 10 dB range. The objective is to determine the appropriate combination of MRC and adaptive filters such that better bit error rate (BER) values are obtained for the given modulation techniques and the channel types. The work considers performance difference in terms of BER rate obtained using both MRC and MRC-equalizer combination. Here, least mean square (LMS), NLMS, and recursive least square (RLS) adaptive filters are taken as part of the setup to determine the best combination for the considered experimental framework. Though the work considers RLS filters, for practical purposes, the LMS-based adaptive filters are preferred due to their overall efficiency [3]. The performance of these approaches are evaluated using semianalytic (SA) and Monte Carlo (MC) approach. The rest of the paper is organized as follows. In Sect. 4.2 the related background is discussed. The experiential details are explained in Sect. 4.3. Results are discussed in Sect. 4.4. The work is concluded in Sect. 4.5.

4.2 Background

One of the common methods preferred for mitigating fading effects is the use of diversity. Among all diversity techniques, MRC is one of the preferred choices. MRC (Fig. 4.5) is a method of diversity combining in which the signals from each channel are added together, the gain of each channel is made proportional to the root mean square (RMS) signal level and inversely proportional to the mean square noise level in that channel different proportionality constants are used for each channel [1].

An adaptive filter self-adjusts its transfer function according to an optimization algorithm driven by an error signal. Because of the complexity of the optimization algorithms, most adaptive filters are digital [3]. Here LMS, NLMS, and RLS filters are used.

Several multipath models have been suggested to explain the observed statistical nature of a wireless channel. Clarke's model based on scattering is one of the most widely preferred and discussed in the literature. Clarke suggested a model which can be used for modeling fading channels in wireless communication [1]. A signal maybe applied to a Rayleigh fading simulator to determine the performance in a wide range of channel conditions. Both flat and frequency-selective fading conditions may be simulated depending upon gain and time-delay settings. The various delayed paths overlap and cause intersymbol interference (ISI). Different frequency components of the transmitted signal experience different fading are known as frequency-selective fading. Delay spread arises from broadband transmission and leads to frequency-selective fading. When the receiver and transmitter are in relative motion with constant speed, the received signal is subjected to a constant frequency shift leading to Doppler spread and produces time-selective fading. It also creates the largest of the frequency shifts of the various paths. Coherent time is defined as the inverse of Doppler spread and uses as a measure of the signal duration at which time-selectivity becomes relevant [1–4].

4.3 System Model and Experimental Considerations

The experimental work carried out can be summarized by the flow logic shown in Fig. 4.1. Initially, the data obtained in binary form are modulated using three different techniques, namely QPSK, BPSK, and DPSK. The aim is to use coherent and noncoherent modulation schemes so as to determine combinations to nullify the effects of stochastic nature of the wireless channel. Another form of the sys-

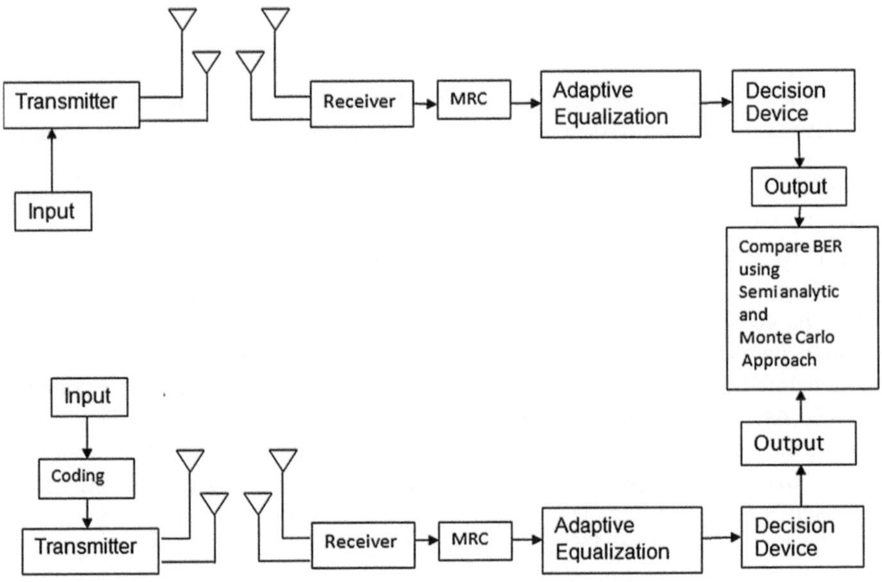

Fig. 4.1 Flow diagram of the system

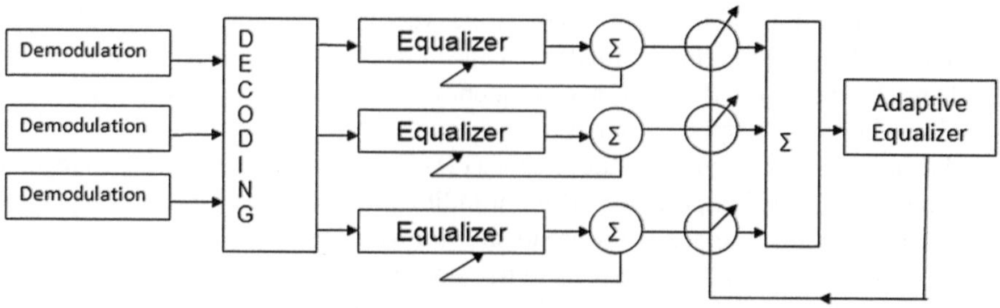

Fig. 4.2 Proposed system model of coding and adaptive equalization assisted MRC

Table 4.1 Parameters used for simulating channel using Clarke-Gans model

Sl. no	Parameter	Value
1	Freq., f_c	900 MHz
2	ω_c	$2\pi f_c$
3	Mobile speed, V	3 kmph
4	No. of paths	8
5	Wavelength, λ	$\frac{3\times 10^8}{f_c}$
6	Doppler shift, f_m	$\frac{V}{\lambda}$
7	Sampling freq., f_s	$8 \times f_m$
8	No. of samples, N	10,000
9	Paths	16
10	Sampling period, T_s	$\frac{1}{f_s}$

tem model is depicted in Fig. 4.2. The signal at the receiver is captured by several diversity branches and passed onto the demodulation blocks. Before demodulation and reconstruction, there is an equalizer block which is combined with the MRC to make improvement in the SNR of the received signal. The equalizer is designed using the LMS algorithm and is provided a mean square error (MSE) convergence goal of 10^{-3}. The equalizer takes around 150 iterations on an average to reach this goal for about 50 sets of data. Four path Rayleigh and Rician channels are generated using the considerations described above. The set of characteristics for multipath fading generated using the parameters given in Table 4.1. Similarly, the Rician channel generated using corresponding considerations yields a set of plots as shown in Fig. 4.3. After the channels are generated, the modulated signals are convolved to provide the transmission effect. Additive Gaussian noise is mixed with values of −3, 1, and 3 dB for three separate channel sets each of size 20. A coded form of the above approach is also proposed. The block diagram is shown in Fig. 4.4. The data before modulation is coded using linear block code (Hamming code) in (7, 4) format which help to improve the quality of the received signal.

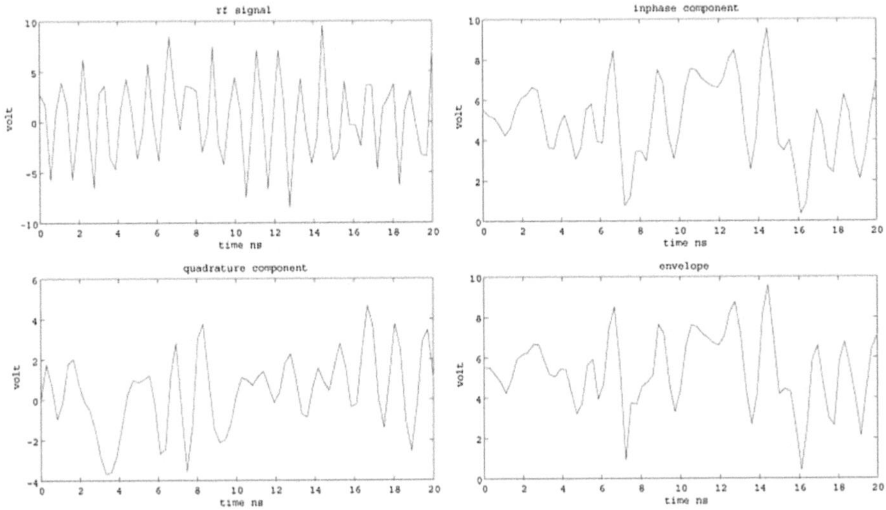

Fig. 4.3 Rician channel path gains

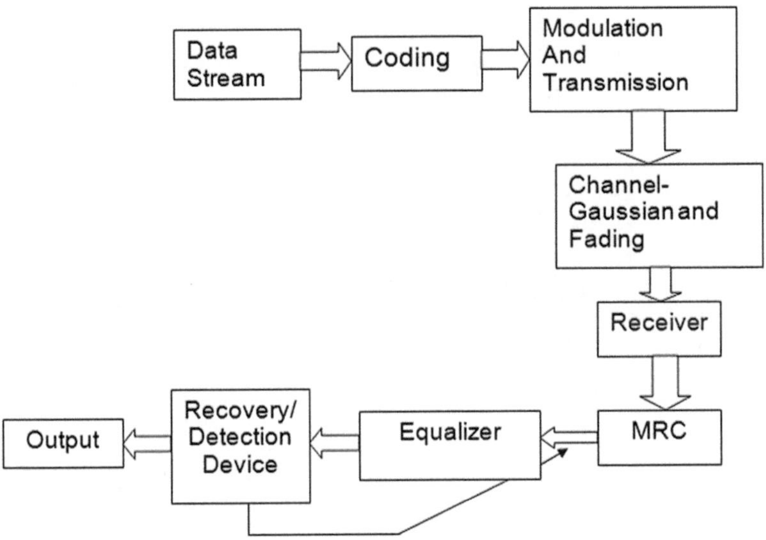

Fig. 4.4 System block diagram showing use of coding

4.4 Results and Discussion

At the receiver end, the MRC process is carried out with two- and three-branch configurations. The three-branch configuration provides better results but at the cost of greater computational constraints. The updating process of the weights of the MRC

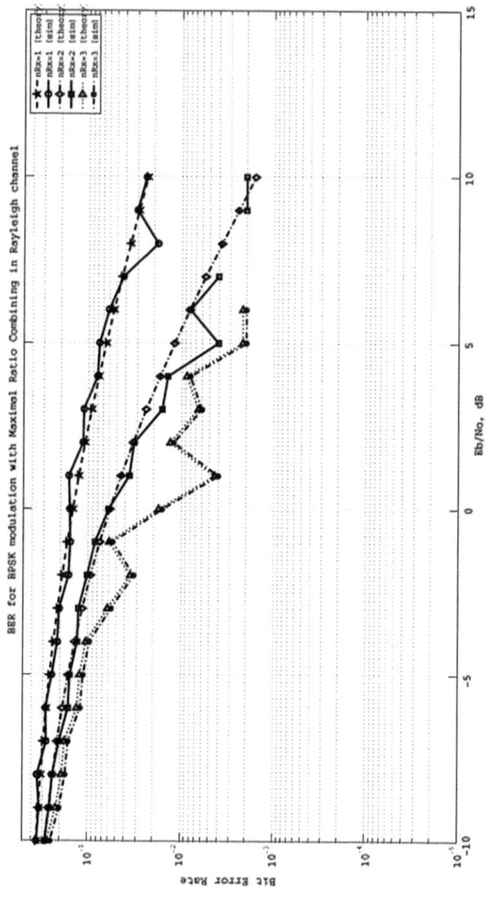

Fig. 4.5 BER plot of MRC with three branches without equalization

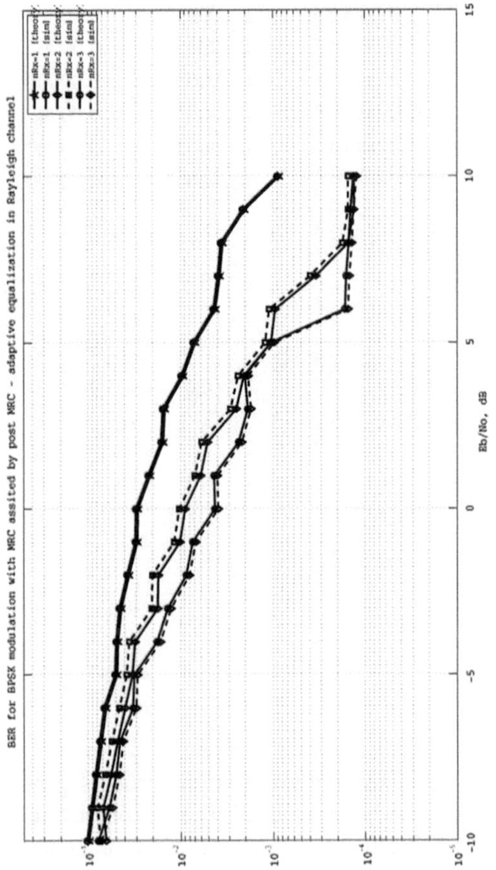

Fig. 4.6 BER plot of MRC with three branches with equalization

Table 4.2 Average MSE of signal inputs at four branches of MRC with coding and adaptive LMS equalization

Sl.no	SNR in dB	MSE for branch MRC wireless channels			
		Branch-1	Branch-2	Branch-3	Branch-4
1	0	0.000071	0.000051	0.000041	0.000049
2	1	0.0000049	0.000051	0.000032	0.000039
3	2	0.000041	0.000048	0.000030	0.000038
4	3	0.000041	0.000040	0.000016	0.000038
5	4	0.000034	0.000036	0.000014	0.000033
6	5	0.000032	0.000035	0.000013	0.000031
7	6	0.000031	0.000033	0.000012	0.000014
8	7	0.000025	0.000031	0.000012	0.000010
9	8	0.000017	0.000032	0.000009	0.000006
10	9	0.000015	0.000025	0.000009	0.000005

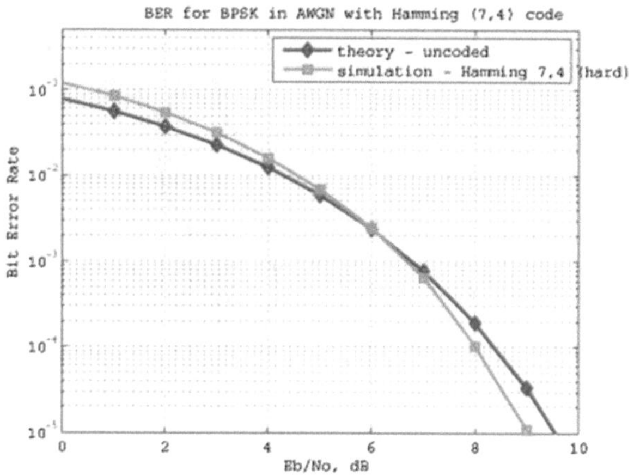

Fig. 4.7 BER calculation by using linear block codes (7, 4)

system is linked to the equalizer in a way that the process continues adaptively till the SNR crosses the fixed threshold value [5]. The use of the equalizer block with the MRC system is thus justified. During training, the MRC branches are supported by the LMS-based adaptive training of the equalizer. Some of the MSE values are shown in Table 4.2. Experiments are carried out with and without the equalizer block. Results are derived from the setup for SNR ranges between −10 and 10 dB. Figure 4.4 shows a BER plot generated for SNR ranges between −10 and 10 dB with three-branch MRC for a Rayleigh multipath fading channel. Figure 4.5 depicts BER of MRC with three branches without equalization. Similarly, a BER plot of MRC with three branches and with equalization is shown in Fig. 4.6. The use of the equalizer block with the MRC system is thus justified. The performance of the system is improved

Table 4.3 Coding gain generated with respect to adaptive equalizer-assisted MRC and coding adaptive equalization with MRC

SNR range (dB)	Uncoded MRC and equalizer (BER)	Coded MRC and equalizer (BER)	Gain
−4	$10^{-1.5}$	$10^{-1.7}$	0.01169
−2	$10^{-2.1}$	$10^{-2.3}$	0.002931
0	$10^{-2.2}$	$10^{-2.5}$	0.0031472
2	$10^{-2.7}$	$10^{-2.9}$	0.0007363
4	$10^{-3.0}$	$10^{-3.2}$	0.0003690
6	$10^{-3.5}$	$10^{-3.8}$	0.00015773
8	$10^{-3.9}$	$10^{-4.3}$	0.00007577
10	$10^{-4.9}$	$10^{-5.3}$	0.00007577

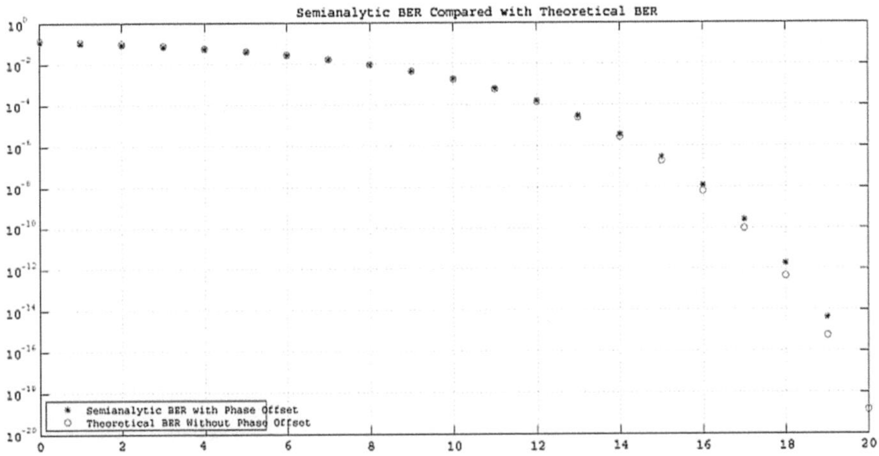

Fig. 4.8 BER calculation using Monte Carlo approach for the three-branch MRC with equalization

further by the use of coding. This is depicted by Fig. 4.7. The use of coding generates certain coding gain. This is summarized in Table 4.3.

The semianalytic (SA) technique works well for certain types of communication systems, but not for others. The SA technique is applicable if a system has all of these characteristics:

- Any effects of multipath fading, quantization, and amplifier nonlinearities must precede the effects of noise in the actual channel being modeled [6].
- The receiver is perfectly synchronized with the carrier. Because phase noise and timing jitter are slow processes, they reduce the applicability of the SA to a communication system.
- The noiseless simulation has no errors in the received signal constellation. Distortions from sources other than noise should be mild enough to keep each signal

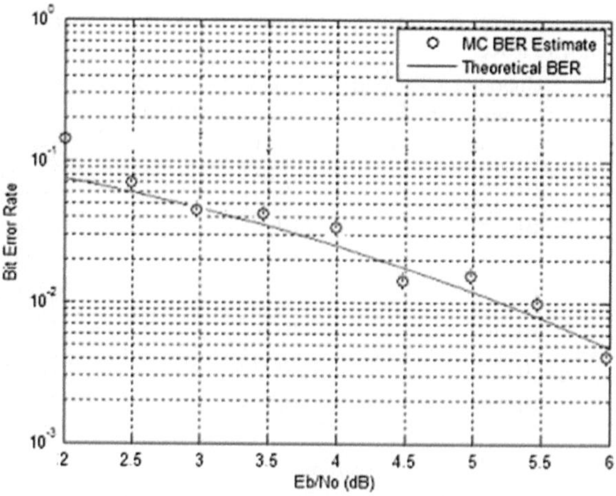

Fig. 4.9 MCS of BER estimation of MRC in Rayleigh faded channel

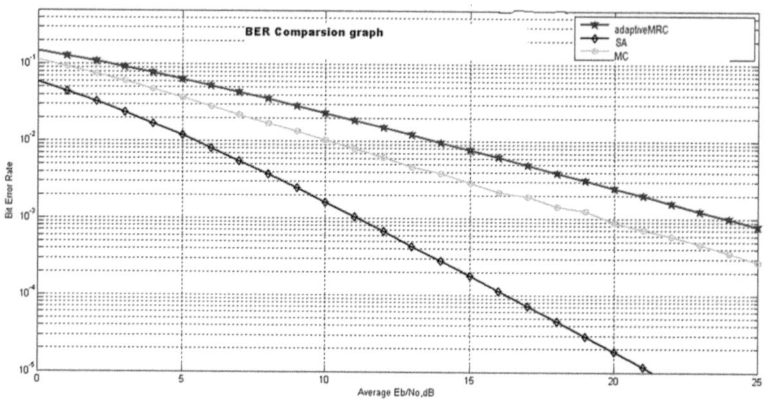

Fig. 4.10 BER comparison graph of adaptive MRC, SA, and MCS

point in its correct decision region. If this is not the case, the calculated BER is too low. For instance, if the modeled system has a phase rotation that places the received signal points outside their proper decision regions, the SA is not suitable to predict system performance.

Monte Carlo Simulation (MCS) is a class of numerical method for computer simulations or computer experiments to generate random sample data based on some known distribution for numerical experiments [7, 8]. A BER calculation by using Monte Carlo approach for the three-branch MRC with equalization is shown in Fig. 4.8. Similarly, MCS of BER estimation of MRC in Rayleigh faded channel is shown in Fig. 4.9. A comparative depiction of adaptive MRC, SA, and MCS is shown in Fig. 4.10. Thus, the use of the coding with adaptive equalization aided MRC improves the performance. This is verified using SA and MCS approaches.

4.5 Conclusion

MRC is an option to mitigate the differences of multipath fading in wireless communications. The work carried out shows that MRC becomes more effective when used in combination with adaptive equalizer. Results shows that at around 10 dB SNR with three-branch MRC, a BPSK modulated signal provides a BER of around 10^{-3} [9]. This value reaches atleast 10^{-4} when the MRC is used in combination an adaptive LMS equalizer. The improvement thus obtained is considerable. The work is further extended to include coded signals. The coding in combination with MRC and adaptive equalization provides further improvement in performance [8]. An evaluation using SA approach and MCS is carried out to obtain a comparative performance measure of MRC adaptive equalization and coding MRC adaptive equalization combinations. The later combination turns out to be superior. The work can be further extended to wireless channels showing frequency selectivity and time variations for applications in mobile networks in rapidly varying environments.

References

1. Rappaport TS (2004) Wireless communications. Principles and practice, 2nd edn. Pearson Education, 7th Indian Reprint, New Delhi
2. Proakis JG (2001) Digital communications, 4th edn. McGraw-Hill Publication, New York
3. Haykins S (2002) Adaptive filter theory, 4th edn. Pearson Education, New Delhi
4. Idris AB, Rahim RFB, Mohd ADB (2006) The effect of additive white gaussian noise and multipath rayleigh fading on BER statistic in digital cellular network. In: Proceedings international conference on RF and microwave conference, Malaysia, pp 97–100
5. Choudhury A, Sarma KK (2011) Adaptive equalization assisted MRC in wireless channels. In: Proceedings of IEEE international conference on information, signal and communication, Gujarat, India
6. Choudhury A, Sarma KK (2011) Coding and adaptive equalization assisted MRC in wireless channels. In: Proceedings of 2nd IEEE national conference on emerging trends and applications in computer science, Shillong, Meghalaya, India
7. Teknomo K (2006) Monte Carlo simulation tutorial. http://people.revoledu.com/kardi/tutorial/Simulation/MonteCarlo.html
8. Choudhury A, Sarma KK (2011) Performance evaluation of coded and adaptive equalization assisted MRC in faded wireless channels using semi analytic and Monte Carlo approach. In: Proceedings of 2nd international conference on advance in computer engineering, Trivandrum, India
9. Berber SM(2001) An automatedmethod forBERcharacteristicsmeasurement. In: Proceedings of 18th IEEE instrumentation and measurement technology conference, vol 3, pp 1491–1495

Chapter 5
ZF- and MMSE Based-Equalizer Design in IEEE 802.15.4a UWB Channel

Tapashi Thakuria and Kandarpa Kumar Sarma

Abstract Ultra wide band (UWB) communication systems occupy huge bandwidths with very low power spectral densities. In a high data rate UWB indoor communication system, the delay spread due to multipath propagation results in inter symbol interference (ISI) which can significantly increase the bit error rate (BER). The distortion and fading caused by the UWB channel and noise sources is removed by equalization which is a signal processing technique. In this work, IEEE 802.15.4a UWB channel model is used for both LOS and NLOS residential environment in a frequency range 2–10 GHz. QAM modulation is used to transmit large volume of data at a time. Equalization is carried out using zero forcing (ZF) and minimum mean square error estimation (MMSE) algorithms. MMSE algorithm shows better performance than the ZF algorithms by reducing BER.

Keywords Ultra wide band (UWB) · Inter symbol interference (ISI) · Bit error rate (BER) · Zero forcing (ZF) · Minimum mean square error estimation (MMSE)

5.1 Introduction

In an ideal communication channel, the received information is identical to that transmitted. However, this is not the case for real communication channels, where signal distortions take place. A channel can interfere with the transmitted data through three types of distorting effects: power degradation and fades, multipath time dispersions, and background thermal noise. Equalization is the process of recovering the data sequence from the corrupted channel samples. Adaptive signal processing and

T. Thakuria (✉)
Department of Electronics and Communication Engineering,
Gauhati University, Guwahati, Assam, India
e-mail: thakuria.tapashi@gmail.com

K.K. Sarma
Department of Electronics and Communication Technology,
Gauhati University, Guwahati 781014, Assam, India
e-mail: kandarpaks@gmail.com

K.K. Sarma et al. (eds.), *Recent Trends in Intelligent and Emerging Systems*,
Signals and Communication Technology, DOI 10.1007/978-81-322-2407-5_5

equalization plays a crucial role in communication systems. It is an important technique to combat distortion and interference in communication links. The conventional approach to communication channel equalization is based on adaptive linear system theory. Channel equalization is a technique used to remove intersymbol interference (ISI) produced due to the limited bandwidth of the transmission channel. When the channel is band limited, symbols transmitted through will be dispersed. This causes previous symbols to interfere with the next symbols, yielding the ISI. Also, multipath reception in wireless communications causes ISI at the receiver. Thus, equalizers are used to make the frequency response of the combined channel equalizer system flat. The purpose of an equalizer is to reduce the ISI as much as possible to maximize the probability of correct decisions [1].

In the near future, indoor communications of any digital data from high-speed to low-speed signals used for timing purposes will be shared over a digital wireless network. Such indoor and home networking is unique, in that it simultaneously requires high data rates (for multiple streams of digital video), low cost (for broad consumer adoption), and low power consumption (for embedding into battery-powered handheld appliances). With its enormous bandwidth, ultra wide band (UWB) provides a promising solution to satisfy these requirements and becomes an attractive candidate for future wireless indoor networks [1]. UWB system is a new wireless technology capable of transmitting data over a wide spectrum of frequency bands for short distances with very low power and high data rates. The UWB technology delivers data rates in excess of 100 Mbps up to 1 Gbps. The UWB not only has potential for carrying high data rate over short distances, but also it can penetrate through doors and other obstacles. The key advantages of the UWB systems over narrowband systems are: high data rate due to the large bandwidth, low equipment cost, low power, and immunity to multipath [2].

UWB communication is a possible technique to achieve higher data rates and better quality, due to its extremely large bandwidth. Performance analysis, such as bit error rate (BER) measure, in a realistic UWB channel is important but a difficult task. This work provides the BER analysis of binary signals under UWB channels. The channel model we consider here is the IEEE 802.15.4a UWB channel and a comparison is made between ZF and MMSE equalizers. In this work we describe the system model and the equalization techniques used.

5.2 Related Works

1. The work [2] shows equalization approaches for high- data-rate transmitted reference (TR) IR-UWB systems employing an autocorrelation receiver front end. Using a maximum likelihood (ML) sequence detector (MLSD) with decision feedback (in the backend of the TR-receiver), a significant improvement of the receiver performance is possible. To avoid high complexity of the MLSD detector, alternative equalizer structures are evaluated. If the parameters of the channel are not known a priori, an equalizer has to be adapted during transmission of train-

ing data. Such an adaptive equalizer is presented as a reference. A linear MMSE equalizer is designed using linear and nonlinear channel coefficients. Then the concept of the linear equalizer is extended to a nonlinear Volterra equalizer which further improves the performance of the IR-UWB receiver structure.

2. A novel equalizer for UWB multiple-input multiple-output (MIMO) channels is characterized by severe delay spreads in [3]. The proposed equalizer is based on reactive tabu search (RTS), which is a heuristic originally designed to obtain approximate solutions to combinatorial optimization problems. The proposed RTS equalizer is shown to perform increasingly better for increasing number of multipath components (MPC), and achieve near ML performance for large number of MPCs at much less complexity than that of the ML detector. The proposed RTS equalizer is shown to perform close to within 0.4 dB of single-input multiple-output AWGN performance.
3. The work [4] presents a new channel shortening algorithm which can shorten the dense multipath channels, such as based on IEEE CM4 model, to just one significant tap. they compare it to maximum shortening signal to noise ratio (MSSNR) and target impulse response (TIR)-based MMSE algorithms and analyze their complexity, consistency, and performance on the basis of number of equalizer taps, captured energy, BER, and signal to interference and noise ratio (SINR) improvement in a multiuser environment. They show that the proposed algorithm is more consistent in shortening the channel with less number of equalizer taps and its performance is well above MSSNR and MMSE algorithms.
4. A robust adaptive MMSE Rake-equalizer receiver [5] scheme is presented with channel estimation that has been transmitted in direct sequence-ultra wideband (DS-UWB) system. The DS-UWB has a fine path resolution by transmitting information with ultrashort pulses. The Rake receiver is known as a technique that can effectively combine paths with different delays and obtain the path diversity gain. Due to sub nanoseconds narrow pulse and large transmission bandwidth in the systems, they use equalizer to overcome the resulting ISI and a long delay spread in the characterization of the UWB channel.
5. In the work [6], a spatial multiplexing single-input multiple-output (SM-SIMO) UWB communication system using the time-reversal (TR) technique is proposed. The system with only one transmit antenna, using spatial multiplexing scheme, can transmit several independent data streams to achieve very high data rate. To cope with the long delay spread of the UWB channel, the TR technique is adopted. TR can mitigate not only the ISI but the multistream interference (MSI) caused by multiplexing several data streams simultaneously as well. Pre-equalization using the channel state information (CSI), which is already available at the transmitter in TR systems, is proposed to further eliminate the ISI and MSI.
6. In the work [7], the authors propose a novel optimization-based pre-equalization filter (PEF) design framework for direct-sequence ultra-wideband (DS-UWB) systems with pre-Rake combining. The key feature in their design is that they explicitly take into account the spectral mask constraints that are usually imposed in practice by the telecommunications regulation and standardization bodies. This avoids the need for an inefficient power back-off, which is necessary for the

existing pre-Rake-based transmitter structures designed solely based on average transmit power constraints.

7. The authors proposed a discrete-time equivalent system model for differential and transmitted reference TR-UWB impulse radio (IR) systems, operating under heavy ISI caused by multipath propagation [8]. In the systems discussed, data are transmitted using differential modulation on a frame-level, i.e., among UWB pulses. Multiple pulses (frames) are used to convey a single bit. Time hopping and amplitude codes are applied for multiuser communications, employing a receiver frontend that consists of a bank of pulse pair correlators.

5.3 Proposed Model for Equalization

As shown in the block diagram in Fig. 5.1, the incoming serial data is first converted from serial to parallel and grouped into x bits each to form a complex number. The number x determines the signal constellation of the corresponding subcarrier, such as 16 QAM. The complex numbers are modulated in the baseband by the inverse FFT (IFFT) and converted back to serial data for transmission. A guard interval is inserted between symbols to avoid ISI caused by multipath distortion. The receiver performs the inverse process of the transmitter. The equalizer is used to correct channel distortion. Here, ZF equalizer is used in IEEE 802.15.3a UWB channel. The performance of ZF and MMSE equalizer is compared in IEEE 802.15.4a channel.

The important blocks of the block diagram shown in Fig. 5.1 are described in the following sections.

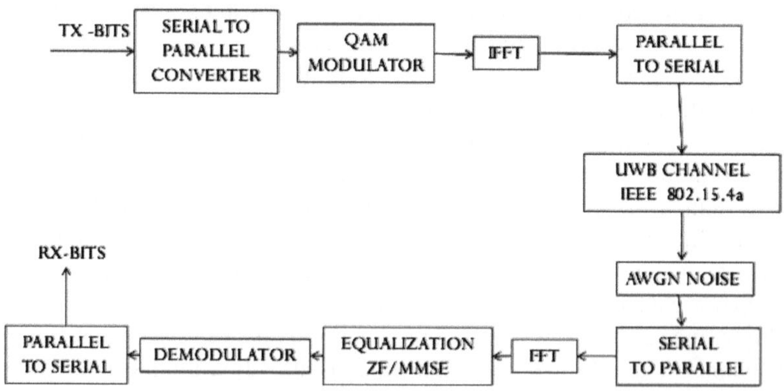

Fig. 5.1 System model for equalization in UWB

5.3.1 OFDM Modulation

In OFDM, the data are divided among a large number of closely spaced carriers. This accounts for the frequency division multiplex (FDM) part of the name. This is not a multiple access technique, since there is no common medium to be shared. The entire bandwidth is filled from a single source of data. Instead of transmitting in serial way, data are transferred in a parallel way. Only a small amount of the data are carried on each carrier, and by this lowering of the bit rate per carrier (not the total bitrate), the influence of ISI is significantly reduced. In principle, many modulation schemes could be used to modulate the data at a low bit rate onto each carrier. Because of dividing an entire channel bandwidth into many narrow subbands, the frequency response over each individual subband is relatively flat [13]. The 'orthogonal' part of the OFDM name indicates that there is a precise mathematical relationship between the frequencies of the carriers in the system. In a normal FDM system, the many carriers are spaced apart in such way that the signals can be received using conventional filters and demodulators. In such receivers, guard bands have to be introduced between the different carriers and the introduction of these guard bands in the frequency domain results in lowering of the spectrum efficiency. It is possible, however, to arrange the carriers in an OFDM signal so that the sidebands of the individual carriers overlap and the signals can still be received without adjacent carrier interference. In order to do this the carriers must be mathematically orthogonal. The receiver acts as a bank of demodulators, translating each carrier down to DC, the resulting signal then being integrated over a symbol period to recover the raw data. Mathematically, suppose we have a set of signals ψ, where ψ_p is the pth element in the set. The signals are orthogonal if

$$\int_a^b \psi_p(t)\psi_q^*(t)dt = 0 \qquad \text{for } p \neq q, \tag{5.1}$$

where the * indicates the complex conjugate and interval [a,b] is a symbol period. At the transmitter, the signal is defined in the frequency domain. It is a sampled digital signal, and it is defined such that the discrete Fourier spectrum exists only at discrete frequencies. Each OFDM carrier corresponds to one element of this discrete Fourier spectrum. The amplitudes and phases of the carriers depend on the data to be transmitted. The data transitions are synchronized at the carriers, and can be processed together, symbol by symbol.

The use of FFT and IFFT in OFDM allows consideration of multiple orthogonal carriers.

5.3.2 QAM Modulation

QAM is both an analog and a digital modulation scheme. It conveys two analog message signals, or two digital bit streams, by changing modulating the amplitudes of two carrier waves, using the amplitude-shift keying (ASK) digital modulation scheme or amplitude modulation (AM) analog modulation scheme. The two carrier waves, usually sinusoids, are out of phase with each other by 90 and are thus called quadrature carriers or quadrature components, hence the name of the scheme. The modulated waves are summed, and the resulting waveform is a combination of both phase-shift keying (PSK) and amplitude-shift keying (ASK), or (in the analog case) of phase modulation (PM) and amplitude modulation [12].

A general expression for the BER of an arbitrary M-ary QAM which is valid not only in AWGN channel but also for any fading channels such as UWB channels is given by the following expression:

$$P_b = \frac{1}{\log_2\sqrt{M}} \sum_{n=0}^{\log_2\sqrt{M}-1} P(\log_2\sqrt{M} - 2n) \tag{5.2}$$

where $P(\log_2\sqrt{M} - 2n)$ denotes the bit error probability of the $(\log_2\sqrt{M} - 2n)$th bit [28].

5.3.3 IEEE 802.15.4a Channel Model

According to measurements of broadband indoor channel, it has been found that amplitudes of multipath fading follow the log-normal or Nakagami distribution rather than the Rayleigh distribution, even if they also show the same phenomenon of clustering as in the Saleh-Valenzuela (S-V) channel model [13]. Based on these results, SG3 a UWB multipath model has been proposed by modifying the S-V model in such away that the multi-cluster signals are subject to independent log-normal fading while the multi-path signals in each cluster are also subject to independent log-normal fading. The ith sample function of a discrete time impulse response in the UWB multipath channel model is given as

$$h_i(t) = X_i \sum \sum a_{r.m}^{(i)} \delta(t - M_m^{(i)} - \tau_{r,m}^{i}) \tag{5.3}$$

where T_m = mth cluster arrival time, $\tau_{r,m}$ = rth ray arrival time in mth cluster. The UWB channel model is different from the S-V channel model in that clusters and rays are subjected to independent log-normal fading rather than Rayleigh fading. More specifically, a channel coefficient is given as

$$\alpha_{r,m} = p_{r,m}\xi_m\beta_{r,m} \tag{5.4}$$

where ξ_m represents log-normal fading of the mth cluster with the variance of σ_1^2 while $\beta_{r,m}$ represents log-normal fading of the rth ray with the variance of σ_2^2 in the mth cluster. The independent fading is assumed for clusters and rays. In Eq. (5.5) $p_{r,m}$ is a binary discrete random variable to represent an arbitrary inversion of the pulse subject to reflection, that is, taking a value of 1 or -1 equally likely. Compared to the channel coefficients of the S-V channel model which has a uniformly distributed phase over $(0, 2\pi)$, those of UWB channel model have the phase of either $-\pi$ or π making the channel coefficient always real. Furthermore, amplitude of each ray is given by a product of the independent log-normal random variables ξ_m and $\beta_{r,m}$. Since a product of the independent log-normal random variables is also a log-normal random variable, a distribution of the channel coefficient $|\xi_m \beta_{r,m}| = 10^{\frac{\mu_{r,m}+z_1+z_2}{20}}$ also follows a log-normal distribution, that is, $20\log_{10}(\xi_m \beta_{r,m}) \sim N(\mu_{r,m}\sigma_1^2 + \sigma_2^2)$, with its average power is given as

$$E[|\xi_m \beta_{r,m}|^2] = \Omega_0 e^{-\frac{T_m}{\tau}} e^{-\frac{\tau_{r,m}}{\gamma}} \tag{5.5}$$

where Ω_0 represents the average power of the first ray in the first cluster. Besides the same set of channel parameters as in the S-V channel model, including the cluster arrival rate Λ, ray arrival rate λ, cluster attenuation constant τ, ray attenuation constant γ, standard deviation σ_x of the overall multipath shadowing with a log-normal distribution, additional channel parameters such as the standard deviations of log-normal shadowing for the clusters and rays, denoted as σ_1 and σ_2, respectively, are required for the UWB channel model. Some proper modifications such as down conversion and filtering are required for implementing the UWB channel in simulation studies, since its bandwidth cannot be limited due to arbitrary arrival times. All the channel characteristics for UWB channel model, including mean excess delay, RMS delay spread, the number of significant paths within 10 dB of peak power (denoted as $\mathrm{NP}_{10\,\mathrm{dB}}$), and PDP, must be determined so as to be consistent with the measurements in practice [13].

It provides models for the following frequency ranges and environments: for UWB channels covering the frequency range from 2 to 10 GHz, it covers indoor residential, indoor office, industrial, outdoor, and open outdoor environments (usually with a distinction between LOS and NLOS properties). For the frequency range from 2 to 6 GHz, it gives a model for body area networks. For the frequency range from 100 to 900 MHz, it gives a model for indoor office-type environments. Finally, for a 1 MHz carrier frequency, a narrowband model is given.

The Task Group 802.15.4a has the mandate to develop an alternative physical layer for sensor networks and similar devices, working with the IEEE 802.15.4 MAC layer. The main goals for this new standard are energy efficient data communications with data rates between 1 kb/s and several Mb/s. Additionally, the capability for geolocation plays an important role. More details about the goals of the task group can be found in the IEEE 802.15.4a PAR. In order to evaluate different forthcoming proposals, channel models are required. The main goal of these channel models is a fair comparison of different proposals. They are not intended to provide information about absolute performance in different environments. Though great efforts have been made to make the models as realistic as possible, the number of available measurements on which the model can be based, both in the 3–10 GHz range, and in the 100–1000 MHz range, is insufficient for that purpose. Furthermore, it was acceptable to do some (over) simplifications that affect the absolute performance, but not the relative behavior of the different proposals. A major challenge for the channel modeling activities is derived from the fact that the PAR and call for proposals does not mandate a specific technology, and not even a specific frequency range. The parameters considered for 802.15.4a residential environment are

PL_0: Pathloss at 1 m distance.
n: Pathloss exponent.
σ_s: Shadowing standard deviation.
A_{ant}: Antenna loss.
κ: Freuency dependence of pathloss.
$\overline{L}$: Mean number of cluster.
λ: Inter cluster arrival rate.
$\lambda_1, \lambda_2, \beta$: Ray arrival rates (mixed poisson model parameters).
Γ: Inter cluster decay constant.
k_γ, γ_0: Inter cluster decay time constant parameter.
$\sigma_{\underline{cluster}}$: Cluster shadowing variance.
$\widehat{m_0}, \widehat{k_m}$: Nakagami m factor variance.
$\widetilde{m_0}$: Nakagami m factor for strong component.
$\gamma_{rise}, \gamma_1, \chi$: Parameters for alternative PDP shape.

The value of the parameters considered for residential environment which is valid upto 20 m are given in Table 5.1.

5.3.4 ZF and MMSE Algorithms for Equalization

Here we briefly discuss the ZF and MMSE algorithms considered for the purpose.

- **Zero forcing (ZF) algorithm** ZF equalizer is a linear equalization algorithm used in communication systems, which inverts the frequency response of the channel. This equalizer was first proposed by Robert Lucky. The ZF equalizer applies the inverse of the channel to the received signal, to restore the signal before the channel. The name ZF corresponds to bringing down the ISI to zero in a noise free case. This will be useful when ISI is significant compared to noise. For a channel with frequency response $F(f)$, the zero forcing equalizer $C(f)$ is constructed such that $C(f) = 1/F(f)$. Thus the combination of channel and equalizer gives a flat frequency response and linear phase $F(f)C(f) = 1$. If the channel response for a particular channel is $H(s)$ then the input signal is multiplied by the reciprocal of impulse response of the channel. This is intended to remove the effect of channel from the received signal, in particular the ISI [14].

- **Minimum mean square error estimator (MMSE) algorithm** An MMSE estimator describes the approach which minimizes the mean square error (MSE), which is a common measure of estimator quality. The main feature of MMSE equalizer is that it does not usually eliminate ISI completely but minimizes the total power of the noise and ISI components in the output. Let x be an unknown random variable, and let y be a known random variable [14]. An estimator $\hat{x}(y)$ is any function of the measurement y, and its mean square error is given by

$$\text{MSE} = E(\hat{X} - X^2) \tag{5.6}$$

where the expectation is taken over both x and y. The MMSE estimator is then defined as the estimator achieving minimal MSE. In many cases, it is not possible to determine a closed form for the MMSE estimator. In these cases, one possibility is to seek the technique minimizing the MSE within a particular class, such as the class of linear estimators. The linear MMSE estimator is the estimator achieving minimum MSE among all estimators of the form $AY + b$. If the measurement Y is a random vector, A is a matrix and b is a vector.

Table 5.1 UWB channel parameter for residential environment

Parameters	LOS	NLOS
Pathloss		
PL_0 (dB)	43.9	48.7
n	1.79	4.58
S (dB)	2.22	3.51
A_{ant}	3 dB	3 dB
κ	1.120.12	1.530.32
Power decay profile		
$\overline{L}$	3	3.5
λ	0.047	0.12
$\lambda_1, \lambda_2, \beta$	1.54, 0.15, 0.095	1.77, 0.15, 0.045
Γ	22.61	26.27
k_γ	0	0
γ_0	12.53	17.50
$\sigma_{cluster}$	2.75	2.93
Small scale fading		
σ_x	0.67	0.69
m_0 (dB)	0	0
k_m	0.28	0.32
m_0 (dB)	0.28	0.32

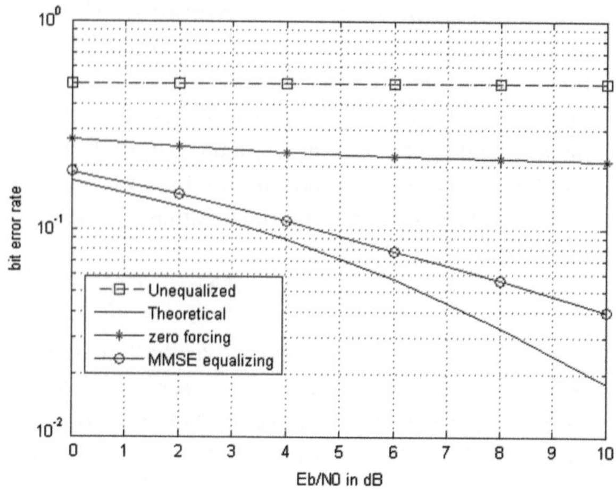

Fig. 5.2 SNR versus BER in LOS environment

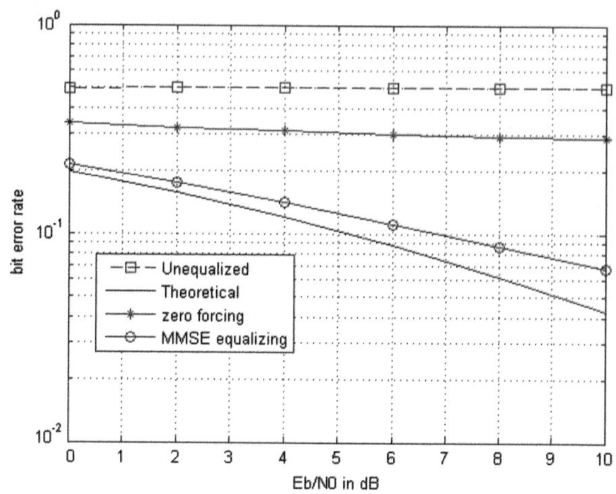

Fig. 5.3 SNR versus BER in NLOS environment

5.4 Result and Discussion

Here 16 QAM modulation is used in OFDM scheme. ZF and MMSE algorithms are used for equalization purpose using both LOS and NLOS residential environments. An SNR range of 0–10 dB is used for plotting the BER curves. One comparison of ZF and MMSE given in these plots is shown in Figs. 5.2 and 5.3. Equation (5.2) is used to generate the theoretical BER curves in all the cases. The ZF approach provides considerable improvements compared to the unequalized case but the MMSE approach shows ever better performance.

The improvement in BER performance obtained using ZF and MMSE approaches is shown in Table 5.2 for both LOS and NLOS cases. There is an improvement of about 64–94 % after the application of ZF and MMSE scheme compared to the

Table 5.2 Improvement in BER in % for ZF and MMSE equalizer

SNR (dB)	Improvement of BER (%)			
	ZF		MMSE	
	LOS	NLOS	LOS	NLOS
0	66	64	67	65
2	74.2	70	69	67
4	77	75.1	74	70
6	79	77	80.5	80
8	81.6	88	80	87
10	82	81	94	91

Table 5.3 Computational time for the system

Number of iteration	Computational time (s)	
	ZF	MMSE
1	14	16
2	25	26
3	39	41
4	48	50

unequalized cases. Also, from Table 5.2 it is seen that the BER improvement is between 67 and 94 % in case of LOS environment in MMSE, for NLOS environment in MMSE it is between 65 and 91 %. The improvement is 66–82 % in case of LOS environment in ZF and 64–88 % in case of NLOS environment. The ZF and MMSE equalizations show improvement with increase in the number of iterations. We have considered upto four number of iterations for both ZF and MMSE cases in LOS and NLOS considerations. We see that for ZF the time required is 14–48 s while in case of MMSE it is between 16 and 50 s. The BER plots shown are for one iteration only which means the worst case performance but with low computational complexity. The performance improves significantly if more number of iterations are considered. But in such a situation the computational efficiency suffers. The computational efficiency can be improved significantly with better hardware design and improved algorithm optimization. Overall, the proposed approach is found to be effective in IEEE 802.15.4a UWB channel for both LOS and NLOS consideration.

5.5 Conclusion

From the experimental results it is seen that MMSE algorithm gives better performance than ZF algorithm in both the UWB channels. The LOS environments of the channels give the improved BER with increasing the SNR value for its lesser complexity than the NLOS environment. The computational time required for MMSE algorithm is slightly more than that of ZF algorithm (Table 5.3).

References

1. Oestges C, Clerckx B (2007) MIMO wireless communication, 1st edn. Elsevier, Oxford
2. Krall C, Witrisal K, Koeppl H, Leus G, Pausini M (2005) Nonlinear equalization for frame-differential IR-UWB receivers. In: International conference on ultra-wideband, pp 576–581
3. Srinidhi N, Mohammed SK, Chockalingam A (2009) A reactive Tabu search based equalizer for severely delay-spread UWB MIMO-ISI channel. In: Proceedings of IEEE GLOBECOM, pp 1–6
4. Husain SI, Choi J (2005) Single correlator based UWB receiver implementation through channel shortening equalizer. Proceedings of Asia-Pacific conference on communications. Perth, Western Australia, pp 610–614
5. Chiu YJ (2009) Adaptive MMSE rake-equalizer receiver design with channel estimation for DS-UWB system. WSEAS Trans. Commun. 8:196–205
6. Nguyen H, Zhao Z, Zheng F, Kaiser T (2010) Preequalizer design for spatial multiplexing SIMO-UWB-TR systems. IEEE Trans. Veh. Technol. 59(8):3798–3805
7. Hamid A, Rad M, Mietzner J, Schober R, Wong VWS (2010) Pre-equalization for pre-rake DS-UWB systems with spectral mask constraints. In: Proceedings of IEEE international conference on communications, pp 1–6
8. Witrisal K, Leus G, Pausini M, Krall C (2005) Channel estimation technique for STBC coded MIMO system with multiple ANN blocks. IEEE J. Sel. Areas Commun. 23(9):1851–1862
9. Fan Z, Timothy OS, Davidson N (2004) Canonical coordinate geometry of precoder and equalizer designs for multichannel communication. In: Proceedings IEEE 5th workshop on sinal procssing advances in wireless communications, pp 401–405
10. Chen IY, Chin WH (2009) Channel spectral flattening in time domain equalizer design for OFDM systems. In: Proceeding of IEEE international conference on communications, pp 1–5
11. Mehrpouyan H (2008) Channel equalizer design based on wiener filter and least mean square algorithms. In: Proceeding of 4th international conference on communication theory, reliability, and quality of service, pp 45–51
12. Rappaport TS, Wireless communications-principle and practice. 2nd edn. Pearson Publication, New Jersey
13. Cho YS, Kim J, Kang CG (2010) MIMO-OFDM wireless communications with matlab. Wiley, Singapore
14. Kumar NS, Shankar KRK (2011) Bit error rate performance analysis of ZF, ML and MMSE equalizers for MIMO wireless communication receiver. Eur. J. Sci. Res. 59(4):522–532
15. Matiae DA, Cimini LJ (1985) Analysis and simulation of a digital mobile channel using orthogonal frequency division multiplexing. IEEE Trans. Commun. 33(7):665–675
16. Welch G, Bishop G (2006) An Introduction to the Kalman Filter. UNC-Chapel Hill, Chapel Hill
17. Sirewongpairat WP, Liu KJR, Ultra wideband communications. Wiley, Hoboken. Available via http://onlinelibrary.wiley.com/doi/10.1002/9780470042397.fmatter/pdf
18. Triki M, Slock DT (2007) Multivariate LP based MMSE-ZF equalizer design consideration and application to multimicrophone dereverberation. IEEE Int. Conf. Acoust. Speech Signal Process. 1:I197–I200
19. Wei C, Gang YC, BiQi L (2008) A low complexity fractionally spaced equalizer design for DTTB system. In: Proceedings of 3rd IEEE conference on industrial electronics and applications, pp 1191–1194
20. Ghani F (2008) Adaptive block linear equalizer for block data transmission systems. Int. J. Eng. Technol. 10(5):25–29

21. Clark MV (1998) Adaptive frequency-domain equalization and diversity combining for broadband wireless communications. IEEE J. Sel. Areas Commun. 16(8):1385–1395
22. Thapar HK, Sands NP, Abbott WL, Cioffi JM (1990) Spectral shaping for peak detection equalization. IEEE Trans. Magn. 26(5):2309–2311
23. Carvalho DB, Filho SN, Searal R (1999) Design of phase equalizers via symmetry of the impulse response. In: Proceedings of the 1999 IEEE international symposium on circuits and systems, pp 37–40
24. Shin DK, Hwang SJ, Sunwoo MH (2000) Design of an DFE equalizer ASIC chip using the MMA algorithm. In: Proceedings of IEEE workshop on signal processing systems, pp 200–209
25. Ma YH, Chen LF, Lee CY (2006) A channel equalizer design for COFDM system. In: Proceedings of international symposium on VLSI design, automation and test, pp 1–4
26. Spinnler B (2010) Equalizer design and complexity for digital coherent receivers. IEEE J. Sel. Top. Quantum Electron. 16(5):1180–1192
27. Comninakis C, Fragouli C, Sayed AH, Wesel RD (1999) Channel estimation and equalization in fading. Conference record of the thirty-third Asilomar conference on signals, systems, and computers, vol 2, pp 1159–1163
28. Najafizadeh L, Tellambura C (2006) BER analysis of arbitrary QAM for MRC diversity with imperfect channel estimation in generalized Rician fading channels. IEEE Trans. Veh. Technol. 55(4):1239–1247

Part II
Selected Issues in Biomedical and Social Science

This section contains three works most of which focus on important issues confronting humanity. A work by Kalita et al. describes the issue of detecting the symptoms of an important problem like child disability using traditional tools and soft computing-based techniques. The work describes how multiple regression and ANN can be effectively used to determine the role of baby's birth symptoms and mother's pregnancy conditions on children's disability. Similarly, the work by Bordoloi et al. deals with the association of Alzheimer's disease with protein structures and the use of ANN for its prediction. In the other work of this section by Goswami et al. the authors have described how ANN can be used for analysis of industrial health.

Chapter 6
Role of Baby's Birth Symptoms and Mother's Pregnancy Conditions on Children's Disability Determined Using Multiple Regression and ANN

Jumi Kalita, Kandarpa Kumar Sarma and Pranita Sarmah

Abstract Birth-cry, birth-weight, mother's distress during pregnancy, baby's health condition soon after birth, are some symptoms that might have some relationship with disability in a child. Influence factors are determined and multiple linear regression and backpropagation artificial neural network (ANN) are applied for modeling the occurrence of disability in a child. Results of multiple regression show that the factors considered have significant effects on the occurrence of disability. Also, the largest beta value (regression coefficient) corresponds to the birth-cry factor of a newborn. It implies the strongest and unique contribution of this variable to explain the dependent variable, which in this case is the proportion of disabled children. An ANN in feedforward form is also configured to perform identical regression for the purpose. Experimental results show that the ANN is a suitable technique for the study of such cases.

Keywords Artificial neural network (ANN) · Multiple regression analysis (MRA) · Multi-layer perceptron (MLP)

J. Kalita (✉)
Department of Statistics, Lalit Chandra Bharali College, Guwahati 781011, India
e-mail: jumikalita@yahoo.co.in

K.K. Sarma
Department of Electronics and Communication Technology, Gauhati University, Guwahati 781014, Assam, India
e-mail: kandarpaks@gmail.com

P. Sarmah
Department of Statistics, Gauhati University, Guwahati 781014, India
e-mail: pranitasarma@gmail.com

K.K. Sarma et al. (eds.), *Recent Trends in Intelligent and Emerging Systems*, Signals and Communication Technology, DOI 10.1007/978-81-322-2407-5_6

6.1 Introduction

Disability in children and the possible factors responsible for disability is a continuing challenge for researchers. The symptoms observed at the birth of a child along with mother's health condition during pregnancy seem to be very important factors that affect disability in children. The aim of this paper is to discuss the effect of four such factors on disability of a child through multiple regression method and artificial neural network (ANN) and to have a comparative analysis of both the methods.

ANNs are nonparametric statistical tools designed using bioinspired approaches [1]. It is a technology for information processing that is inspired from human brain and nervous systems. ANNs are designed for use in many areas such as developing models, time series analysis, pattern recognition, signal processing and control, etc. It can learn input patterns and use this knowledge to make predictions. ANNs are known to be adaptive, robust, and can learn process data irrespective of methods of creation. Such an approach has been applied in this paper along with the conventional statistical method of multiple linear regression to study the effect of birth cry, birth weight, baby's problem within a short period after birth, and mother's health condition during pregnancy on disability of a child.

6.2 Review of Literature

Several authors have compared performances of ANN and multiple regression, on areas ranging from economics to medical science.

Moradzadehfard et al. [2] compared the predictive power of neural networks and regression model to estimate dividend payout ratio and found neural network to be more accurate.

ANN performed better than multiple regression analysis (MRA) in the predictive performance of housing sales [3]. The authors used moderate to large size of data sample.

Kim et al. [4] proposed a model in their study to estimate systolic BP (SBP) conveniently and indirectly using PTT and some biometric parameters such as weight, height, body mass index (BMI), length of arm, and circumference of arm. Based on the above data, they found that ANN had better accuracy than multiple regression in estimating BP, and ANN's estimation satisfied AAMI standard as a BP device.

Pugh and Ryman [5] used neural network and multiple regression in their study for a medical diagnosis. They found that for large samples neural networks could be useful. For small data sets they found that neural network algorithms tend to overfit on the validation samples and displayed a relatively large degree of shrinkage upon cross-validation. Analysis of data with a nonlinear component demonstrated the ability of the neural network to fit either a polynomial or interactive term without the user having to mode such terms. They also found that if these effects were modeled, the regression equation performed well.

Subramanian et al. [6] used ANN and multiple regression analysis using 33 factorial design (FD) to optimize the formulation parameters of cytarabine liposomes and found that ANN provides more accurate prediction and is quite useful in the optimization of pharmaceutical formulations compared with the multiple regression analysis method.

Pao [7] used multiple linear regression and ANN models to analyze the important determinants of capital structure with seven explanatory variables of corporation's feature and three external macro-economic control variables, of high-tech and traditional industries in Taiwan and they found that the ANN models achieve better fit and forecast than regression models for debt ratio (based on the values of root mean square error) and ANNs are capable of catching sophisticated nonlinear integrating effects in both types of industries.

Chi and Wang [8] in their study of airport visibility found statistically significant influence factors like humidity, visual hazard, temperature, and atmospheric condition as causes for low airport visibility. They used multiple regression and ANN for this purpose. They also found backpropagation artificial neural network approach to be superior to the conventional regression method by comparing the coefficients of determination.

In general, the literature supports use of ANN over MRA on a broad spectrum of problems. However, suitability of these two methods has not been tested on disability data. So, both these methods have been applied to predict disability in children based on the assumed factors mentioned above.

6.3 Data Source and Methodology

A total of 283 medically verified data sets were collected from the Guwahati centre of national institute of public cooperation and child development (NIPCCD), India for this study.

6.3.1 Multiple Linear Regression

Multiple linear regression attempts to model the relationship between two or more explanatory variables and a response variable by fitting a linear equation to observed data.

For 'k' independent variables $x_1, x_2, \ldots, x_k$ and n observations $y_1, y_2, \ldots, y_n$ each of which can be expressed by the equation

$$Y_i = \beta_0 + \beta_1 X_{1i} + \beta_2 X_{2i} + \beta_3 X_{3i} + \beta_k X_{ki} + \varepsilon_i \tag{6.1}$$

for, i =1, 2,…, n;

where,

$$y = \begin{pmatrix} y(1) \\ y(2) \\ y(3) \\ . \\ . \\ y(\mathrm{n}) \end{pmatrix}_{n\times 1} \tag{6.2}$$

$$X = \begin{pmatrix} 1 & x_{11} & x_{12} & . & . & x_{1k} \\ 1 & x_{21} & x_{22} & . & . & x_{2k} \\ . & . & . & . & . & . \\ . & . & . & . & . & . \\ 1 & x_{n1} & x_{n2} & . & . & x_{nk} \end{pmatrix}_{n\times p} \tag{6.3}$$

$$\beta = \begin{pmatrix} \beta(1) \\ \beta(2) \\ \beta(3) \\ . \\ . \\ \beta(\mathrm{k}) \end{pmatrix}_{p\times 1} \tag{6.4}$$

$$\varepsilon = \begin{pmatrix} \varepsilon(1) \\ \varepsilon(2) \\ \varepsilon(3) \\ . \\ . \\ \varepsilon(\mathrm{n}) \end{pmatrix}_{n\times 1} \tag{6.5}$$

and where, $p = k + 1$ with $\varepsilon\ N(0, \sigma^2)$.
Then the least squares solution for estimation of β involves finding b for which the sum of squares of errors (SSE) is minimized. This minimization process involves solving b in the equation

$$\frac{\delta}{\delta b}(\mathrm{SSE}) = 0 \tag{6.6}$$

where,

$$\mathrm{SSE} = (y - Xb)'(y - Xb) \tag{6.7}$$

The result reduces to the solution of b in

$$(X'X)b = X'y \tag{6.8}$$

$$A = X'X = \begin{pmatrix} n & \sum_{i=1}^{n} x_{i1} & \sum_{i=1}^{n} x_{i2} & \cdot\cdot & \sum_{i=1}^{n} x_{ik} \\ \sum_{i=1}^{n} x_{i1} & \sum_{i=1}^{n} (x_{i1})^2 & \sum_{i=1}^{n} x_{i1}x_{i2} & \cdot\cdot & \sum_{i=1}^{n} x_{i1}x_{ik} \\ \cdot & \cdot & \cdot & \cdot\cdot & \cdot \\ \cdot & \cdot & \cdot & \cdot\cdot & \cdot \\ \sum_{i=1}^{n} x_{ik} & \sum_{i=1}^{n} x_{ik}x_{i1} & \sum_{i=1}^{n} x_{ik}x_{i2} & \cdot\cdot & \sum_{i=1}^{n} x_{ik}x_{ik} \end{pmatrix}_{1\times 2} \tag{6.9}$$

$$X'y = \begin{pmatrix} \sum_{i=1}^{n} y_i \\ \sum_{i=1}^{n} x_{i1} y_i \\ \cdot \\ \cdot \\ \sum_{i=1}^{n} x_{ik} y_i \end{pmatrix} \tag{6.10}$$

If the matrix A is nonsingular, we can write the solution for the regression coefficients as

$$b = (X'X)^{-1}X'y \tag{6.11}$$

The sum of squares identity is-

$$\sum_{i=1}^{n}(y_i - \bar{y}^2 = \sum_{i=1}^{n}(\hat{y}_i - \bar{y}^2 + \sum_{i=1}^{n}(y_i - \hat{y}^2) \tag{6.12}$$

i.e., Total sum of squares (SST) = Regression sum of squares (SSR) + Error sum of squares (SSE).
In matrix notation, this is expressed in Eq. 6.13.

$$y'y = b'X'y + (y'y - b'X'y) = y'X(XX)^{-1}X'y + [y'y - y'X(X'X)^{-1}X'y] \tag{6.13}$$

The analysis of variance leads to a test of

$$H0 : \beta_1 = \beta_2 = \beta_3 = ... = \beta_k = 0$$

The test for significance of regression is a test to determine if there is a linear relationship between the response y and any of the regressor variables $x_1, x_2, ..., x_k$. The rejection of the above null hypothesis implies that at least one of regressors contributes significantly to the model.

6.3.2 Artificial Neural Network

An ANN is a mathematical or computational model that is inspired by the structure and/or functional aspects of biological neural networks. It consists of an interconnected group of artificial neurons, and it processes information using a connectionist approach to computation. In most cases an ANN is an adaptive system that changes its structure based on external or internal information that flows through the network during the learning phase. Modern neural networks are nonlinear statistical data modeling tools. They are usually used to model complex relationships between inputs and outputs or to find patterns in data. ANNs can be trained to perform complex functions in various fields, including pattern recognition, identification, classification, speech, vision, and control system and to solve problems that are difficult for conventional computers or human beings [1]. Depending on various learning rules, the following fundamental classes of ANN architecture are defined as mentioned below:

- Single-layer feed-forward network: In layered neural network the neurons are organized in the form of layers. The single layer feed forward network has only one layer.
- Multi-layer feed-forward network: It has one or more hidden layers between the input and output layers, whose computation nodes are called hidden neurons or hidden units. The function of neurons is to intervene between the external input and the network output in some useful manner. By adding one or more hidden layers the network is enabled to extract higher order statistics. The ability of hidden neurons to extract higher order statistics is particularly valuable when the size of the input layer is large. A multi-layer perceptron (MLP) consists of multiple layers of nodes in a directed graph, with each layer fully connected to the next one (Fig. 6.1). Except for the input nodes, each node is a neuron (or processing element) with a nonlinear activation function. MLP utilizes a supervised learning technique called backpropagation for training the network [1].

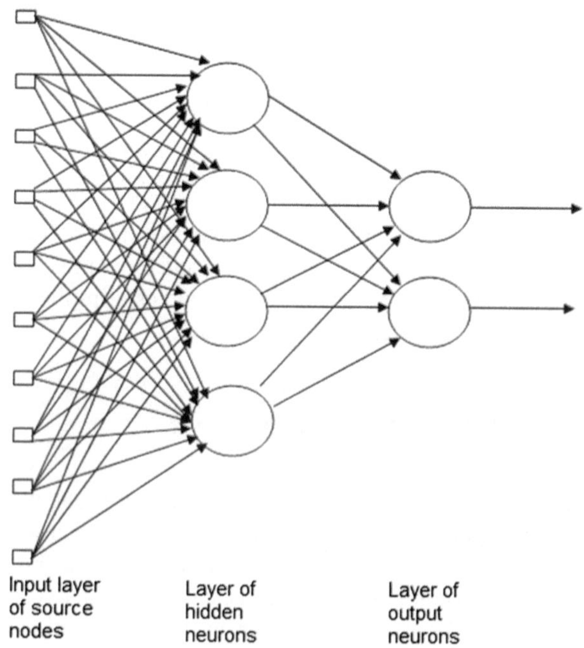

Fig. 6.1 Multi-layer feed-forward network

6.4 Experimental Result and Discussion

6.4.1 Multiple Regression

For this study the indicator variables X_1, X_2, X_3 and X_4 are defined as follows:

$$X_1 = \begin{cases} 0, \text{ if birth is absent;} \\ 1, \text{ otherwise.} \end{cases} \tag{6.14}$$

$$X_2 = \begin{cases} 0, \text{ if birth weigth is less than 2.5 kg;} \\ 1, \text{ otherwise.} \end{cases} \tag{6.15}$$

$$X_3 = \begin{cases} 0, \text{ for babies having problem soon after birth;} \\ 1, \text{ otherwise.} \end{cases} \tag{6.16}$$

$$X_4 = \begin{cases} 0, \text{ for mothers having health problem during pregnancy;} \\ 1, \text{ otherwise.} \end{cases} \tag{6.17}$$

There are sixteen different combinations of the indicator variable are present. The regression of Y on X_1, X_2, X_3 and X_4 is given by

$$Y_i = \beta_0 + \beta_1 X_{i1} + \beta_2 X_{i2} + \beta_3 X_{i3} + \beta_4 X_{i4} + \varepsilon_i \tag{6.18}$$

where, i = 1, 2,...,16
Equivalently,

$$Y = X\beta + \varepsilon \tag{6.19}$$

where, $Y = (Y_1, Y_2, \ldots, Y_{16})'$ $\beta = (\beta_0, \beta_1, \beta_2, \beta_3, \beta_4)'$
and $X = (X_{ij})_{16*4}$ and $\varepsilon = (\varepsilon_i)_{16*4}$ with ε $N(0, \sigma^2)$

In this study the null hypothesis to be tested is: H0: All the regression coefficients are zero.

The data set has been analyzed using SPSS and the results are summarized in Tables 6.1, 6.2 and 6.3.

R^2 is 0.797 implies that the model explains 79.7 % of the variance in the dependent variable, i.e., the proportion of disabled children.

Table 6.1 Results of multiple regression[a]

Model	R	R^2	Adjusted R^2	Standard error of the estimate
1	0.893[b]	0.797	0.724	0.0193712

[a] Dependent variable: proportion of disability
[b] Predictors: (constant), Mother's problem, Baby's problem, Birth weight, Birth cry

Table 6.2 Results of ANOVA[a]

Model	Sources of variation	Sum of squares	df	Mean square	F	Significance
1	Regression	0.016	4	0.004	10.820	0.001[b]
	Residual	0.004	11	0.000		
	Total	0.020	15			

[a] Dependent variable: proportion of disability
[b] Predictors: (constant), Mother's problem, Baby's problem, Birth weight, Birth cry

Table 6.3 Multiple regressions coefficients[a]

	Model	Unstandardized coefficients		Standardized coefficients	t	Significance	Collinearity statistics	
		B	Standard error	Beta			Tolerance	VIF
1	(Constant)	0.054	0.011		4.953	0.000		
	Birth cry	0.052	0.010	0.733	5.400	0.000	1.000	1.000
	Birth weight	−0.023	0.010	−0.323	−2.380	0.037	1.000	1.000
	Baby's problem	0.013	0.010	0.186	1.373	0.197	1.000	1.000
	Mother's problem	−0.025	0.010	−0.348	−2.563	0.026	1.000	1.000

[a] Dependent variable: proportion of disability

The ANOVA table with significance p value $0.001 < 0.01$ implies that factors considered have significant effects on the occurrence of disability (rejecting the null hypothesis and concluding that at least one of the regression coefficient is different from zero).

From the standard coefficient[1] (converted to the same scale) column of the coefficient box above the largest beta value 0.733 corresponds to birth cry of a newborn. It implies that this variable makes the strongest unique contribution to explain the dependent variable.

Also, there exists no multicollinearity among predictor variables which is clear from the column headed by tolerance of the table labelled coefficients above.

[1] Standard coefficients are used to compare the different variables. These values for each of the different variables have been converted to the same scale. Unstandardized coefficient values are used in construction of regression equation. Tolerance is calculated by 1-R2 for each variable. A very lower tolerance value (near 0) indicates the possibility of multicollinearity.

6.4.2 Artificial Neural Network

For this study, a coding scheme of the input sample data is developed using Boolean logic so that the patterns fed to the ANN are sequences of 1 and 0 s. Four critical symptoms are identified and placed along the columns of a table. Next, from the collected data a coded input data set is formed depending on the symptom observed. If a specific symptom is present 1 is assigned to the relevant column, while a 0 is placed if absent. After the encoding process of the complete data set is ready two broad classes of pattern ensembles are formed. The first set is used for training the ANN only. The other data set is used to check the level of training of the ANN and is known as validation set. Then another set is formed to check the ability of the ANN extensively and make inferences regarding its ability to provide prediction with variations in input patterns. Along with these the ANN specifications given in Table 6.4 are used for this study. The results obtained are summarized graphically in Fig. 6.2.

Table 6.4 ANN specifications

Sl. no.	Item	Description
1	ANN type	Multi layer perceptron
2	Number of hidden layers	Two
3	Activation functions in the hidden layers	'Tansig' and 'logsig'
4	Training method	Gradient descent with adaptive learning rate
5	Epochs	20–200
6	Maximum training time	2.9 s
7	Learning rate	0.1–0.5
8	Number of trials	Atleast 10 for each epoch case
9	Average success rate range	50–98 %
10	MSE goal	0.0001

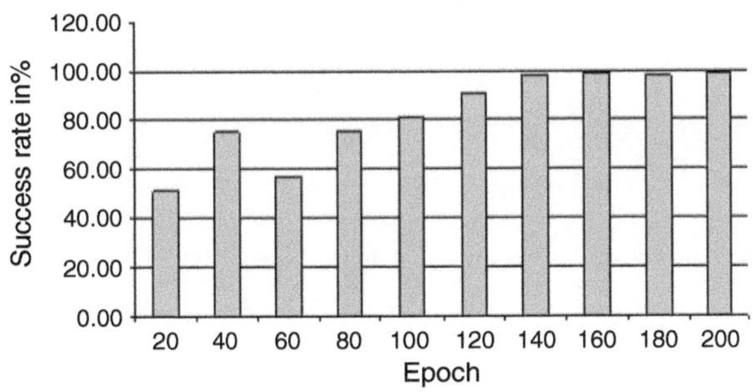

Fig. 6.2 Average success rate versus epoch obtained from MLP training

6.5 Conclusion

The collected data on children with disabilities are subjected with statistical regression method to determine the effect of some symptoms (birth-cry, birth-weight, mother's distress during pregnancy, baby's health condition soon after birth) that appear at the time of birth of a baby and during pregnancy of the mother on the extent of disability in children. The average success rate achieved is around 80 %. The same data are analyzed using ANN in feed-forward configuration. The ANN in multi-layer perceptron (MLP) form is trained with error backpropagation algorithm for a number of epochs. The success rates are found to be between 51 and 98 %. Both the methods are suitable for analyzing disability data, but ANN yields higher success rate in the prediction if number of epoch is increased.

References

1. Haykin S (2006) Neural networks, 2nd edn. Dorling Kindersley, India
2. Moradzadehfard M, Motlagh PA, Fathi MR (2011) Comparing neural network and multiple regression models to estimate dividend payout ratio. Mid East J Secur. Resolut 10(3):302–309 (IDOSI Publications)
3. Nguyen N, Cripps A (2001) Predicting housing value: a comparison of multiple regression analysis and artificial neural networks. JRER 22(3):313–336
4. Kim JY, Cho BH, Im SM, Jeon MJ, Kim IY, Kim SI (2005) Comparative study on artificial neural network with multiple regressions for continuous estimation of blood pressure. In: Proceedings IEEE 27th annual conference engineering in medicine and biology. Shanghai, China, pp 1–4
5. Pugh, WM, Ryman DH (1992) A comparison of multiple regression and a neural network for predicting a medical diagnosis. In: Interim report, Jul–Sep 91, Naval Health Research Center, San Diego, CA
6. Subramanian N, Yajnik A, Murthy RSR (2004) Artificial neural network as an alternative to multiple regression analysis in optimizing formulation parameters of cytarabine liposomes. AAPS PharmSciTech 5(1). http://www.aapspharmscitech.org
7. Pao HT (2008) A comparison of neural network and multiple regression analysis in modeling capital structure. Expert Syst Appl Elsevier 35:720–727
8. Chi TH, Wang YM (2011) Using multiple regression and artificial neural network approach for modeling airport visibility. In: Proceedings of international conference on agriculture and biosystems engineering. Adv Biomed Eng 1–2:428–431

Chapter 7
A Soft Computational Framework to Predict Alzheimer's Disease (AD) from Protein Structure

Hemashree Bordoloi and Kandarpa Kumar Sarma

Abstract Alzheimer's disease is a common disease which is characterized by a person losing his memory progressively. Finally, the person also loses his life. It is often seen in the people above the age 60 but it may occur early. This disease destroys memory cells of the brain. Till now, it is a disease without any treatment and also there are no proper means of diagnosis. Research shows that most often it occurs either due to the deposition of defective structure of amyloid protein or due to the tangles in the brain. In this paper, we have proposed a system to detect the defective Amyloid protein using two classifiers. Secondary structure of Amyloid protein is detected and analyzed in our work which provides a way to predict the cause of Alzheimer.

Keywords Alzheimer's disease · Amyloid protein · Amino acids · Artificial neural network (ANN) · Multilayer perceptron (MLP)

7.1 Introduction

Alzheimer's disease (AD) is an illness of the human brain that progressively increase the ill effects. Basically, memory loss and improper functioning of brain are the two primary symptoms of AD. Since memory loss is related to AD, it is termed as one of the most common form of dementia. It is categorized under progressive disease category because dementia symptoms gradually worsen over a number of

H. Bordoloi (✉)
Department of Electronics and Communication Engineering,
Assam Don Bosco University, Guwahati, Assam, India
e-mail: hbmaini@gmail.com

K.K. Sarma
Department of Electronics and Communication Technology,
Gauhati University, Guwahati 781014, Assam, India
e-mail: kandarpaks@gmail.com

K.K. Sarma et al. (eds.), *Recent Trends in Intelligent and Emerging Systems*,
Signals and Communication Technology, DOI 10.1007/978-81-322-2407-5_7

years. In its early stage, memory loss occurs but in its last stage a person suffering from this disease lost his conversation ability and they become unable to respond to environment and finally leads to death [1].

Till now, causes of AD is not properly detected but according to medical researchers, basically genetic, lifestyle, and environmental factors affect the brain in AD. But the effect of AD is properly analyzed. It damages and kills brain cells. An AD affected brain has a few numbers of connections with the surviving cells than a healthy brain.

Under microscope, two types of abnormalities can be seen in the cells of an AD effected brain. The first one is plaques. These are clumps of a protein called beta amyloid. Deposition of beta amyloid on the outside of the brain cell is considered to be a prime suspect for AD. It may damage and destroy brain cells. It can also interfere in cell-to-cell communication. The second are the tangles. These occur due to the abnormal twist of a protein called Tau protein. This effect in the proper functioning of brain transport system used to carry nutrients and other essential materials throughout their long extensions [1, 2].

Certain changes are seen in case of AD affected person. For example, steady decline in memory and mental function, disorientation and misinterpreting of spatial relationships, problem to identify right words and problem in conversation with other persons, loss of reasoning and thinking capacity, etc. Here, we propose a system to predict AD by analyzing the amyloid protein using soft computational tools.

7.2 Related Works

Alzheimer's disease was first described by a German psychiatrist and neuropathologist Alois Alzheimer in 1906 and was named after him [1]. During the last two decades, several works have been completed on AD which deals with detection of different factors of AD. It is seen that most of the work done previously are related with DNA and chromosomal disorder. But as per our knowledge, there is no such work has been done based on the chemical structures of the proteins responsible for AD. In our present work, we are exploring the possibility of AD based on protein structure. Before the work, a detailed review of the related literature is required. We include here a few relevant works reported in current literature to follow the different steps used in the following work done.

In the Royal Perth hospital of Australia, AD is analyzed by immunocytochemical methods upon few numbers of aged brains. As compared to other histological technique, immunocytochemical technique is more sensitive and allows an easier detection between normal and abnormal brain [2].

In Imperial college of Science, Technology and Medicine, London, a work is done to detect AD considering different concentration of amyloid protein. Some parametric tests are used to explore concentration differences [3].

In another approach, plasma biomarker for AD is identified by using baseline plasma screening technique. In this technique, 151 multiplexed analysts are combined with targeted biomarker and clinical pathology data are used [4].

A neuro imaging technique has been employed for detection of AD in one approach. Images of affected brains are analyzed in this technique. Here, PiB PET imaging technique is used to detect deposition of beta Amyloid protein in brain cells [5].

7.3 Proposed Work

Till now, amyloid beta protein is considered as biomarker of Alzheimer's disease. Amyloid is a class of protein which share several properties. They have beta-pleated structure with high affinity to bind congo red. In polarized light, birefringence is produced by this protein. They can also produce an X-ray diffraction pattern. They are insoluble protein with unrelated amino acid sequences [6].

Presence of amyloid protein shows incompatibility with normal cell function and survival. These proteins destroy the cell function and leads to the death of cell in which they form [6].

From the detailed literature survey, we have seen that till now researches are done either by considering images of brain or by the concentration of Amyloid protein. But from the literature survey, we see that no work is reported by considering the structure of amyloid protein. That is why, we have proposed a simulation-based model which deals with the abnormality of the structure of amyloid protein. In our approach, we have first considered the chemical components of the primary structure of amyloid protein and then we analyzed secondary structure of the same.

The primary structure of Amyloid protein is given as—DAEFRHDSGYEVHHQK LVFFAEDVGSNKGAIIGLMVGGVVIA. This is the primary structure of amyloid protein in case of normal brain. Here, each letter signifies amino acids. For example, D means aspartic acid, A means alanine, and so on. In case of AD effected brain, primary structure of amyloid protein is changed. In that case, some amino acids in the sequence are substituted by some other amino acids. The amino acid sequence of Amyloid in case of AD affected brain is DAEFAADSGYEVAAQALVFFAED-VGSNKGAIIGLMVGGVVIA. Substitution that occurs in AD effected brain is Alanine replaces Histidine at position 6, 13, 14 Lysine at position 16 and Arginine at position 5. In our work, we have designed a classifier which classifies secondary structure of Amyloid protein based on the primary structure. In our work, we have used artificial neural network (ANN) as classifier. It is a bio-inspired, simulated distributed processor. They are made up of simple processing units called artificial neurons that have a natural propensity for storing experimental knowledge and making it available for use. It can acquire knowledge from its environment through a learning process. To acquire knowledge, synaptic weights are used [7].

7.4 Experimental Setup

The experimental setup of our proposed work consists of two ANN classifiers. First classifier classifies primary structure of amyloid protein based on the amino acids sequence. Then output of the first classifier is given as the input to the second classifier which detects the secondary structure of Amyloid protein. The system model for our proposed work is shown in Fig. 7.1. Our work follows different steps as described below.

7.4.1 Data Set Selection

Amyloid protein is the biomarker for AD. So we have considered primary structure of Amyloid protein as our data. Primary structure of Amyloid protein is as follows: DAEFRHDSGYEVHHQKLVFFAEDVGSNKGAIIGLMVGGVVIA, where each letter signifies amino acid.

For AD affected brain this sequence is changed as follows: DAEFAADSGYEVAAQALVFFAEDVGSNKGAIIGLMVGGVVIA.

These two sequences are used as data.

7.4.2 Generation of Coding Scheme

The amino acid sequences are coded based on a coding scheme. This coding scheme is generated based on the chemical structure of each amino acid.

7.4.3 Configuration of ANN

In our work, we have used ANN as classifier. To update the weight of the network, Backpropagation algorithm is used. Two fully connected multilayer perceptron (MLP), feedforward network is used to classify amino acids. In our classifier, one hidden layer is used. First, classifier detects primary structure which is used as input

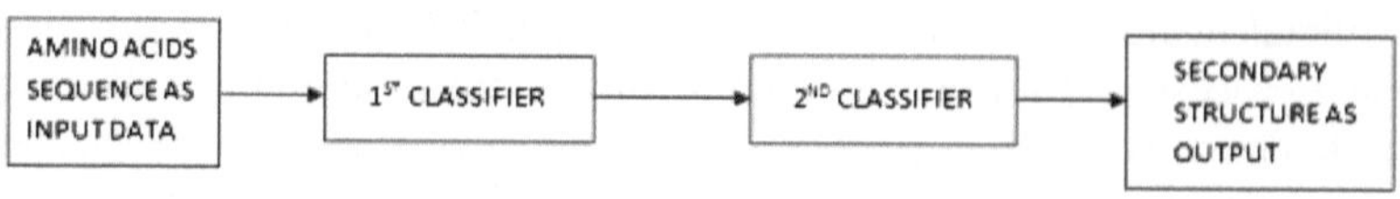

Fig. 7.1 Proposed system model

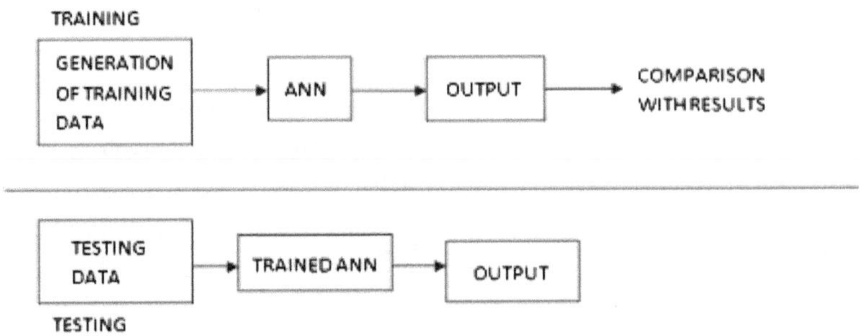

Fig. 7.2 Proposed model showing training and testing phases of classifier

to the second classifier. Second, classifier detects secondary structure of amyloid protein.

7.4.4 *Training and Testing of ANN*

The network is trained with normal and abnormal structures of amyloid protein. With the coded data, testing is done. 1st classifier is trained with chemical structure of Amyloid protein which identifies primary structure of this protein. The 2nd classifier is then trained with the primary structure which gives the desired result as secondary structure of amyloid protein with AD and without AD. The proposed model composed of two classifiers as shown in Fig. 7.2. The system is formed by two-level classifiers. Amino acids sequences are used as inputs for the first-level classifier which provides the identification of the primary structure. This primary structure is then applied to the second-level classifiers that predict secondary structure of Amyloid protein.

The entire work may be summarized by the following steps:

- Extraction and coding of the primary structure based on the chemical structure
- Training and testing of the protein structure and
- Derivation of secondary structure.

7.5 Results

The ANN is trained with normal as well as abnormal primary structure of amyloid protein. Training is carried out using error back propagation algorithm with Levenberg-Marquardt optimization. Training and testing of both the networks are done in the same method. During training, 42 data are given to ANN to handle. The required parameters are recognized with 100 % accuracy. Performance goal of 10^{-6}

Table 7.1 Parameters of proposed ANN

ANN	MLP
Data set size	Training=42; Testing=35
Training type	Back propagation with Levenberg-Marquardt optimization
Maximum no. of epochs	3500
Variance in training data	50 %

Table 7.2 Performance of ANN set up

Epoch	Time (s)	MSE	Rate (%)
1200	30	10^{-2}	92
1500	45	10^{-3}	94
2000	67	10^{-5}	97
2500	78	10^{-6}	100

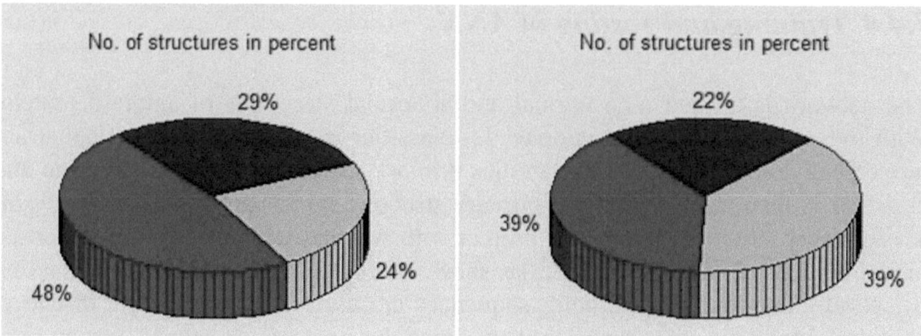

Fig. 7.3 Amyloid protein without AD and amyloid protein with AD respectively

is given to the ANN and it is attained in around 78 s. The network is trained with 42 samples and tested with 35 samples as shown in Table 7.1. Again Table 7.2 clearly indicates that the proposed ANN shows nearly 100 % accuracy under best training conditions with a MSE convergence of about 10^{-6}. By varying the training time, we are able to make our classifier robust enough. The 2nd network is also trained and tested with the same method as that of the first one. Combination of both systems gives a satisfactory result. The Fig. 7.3 represents the performance of ANN classifier showing the secondary structure. As shown in the Fig. 7.3, in case of normal structure of Amyloid protein, it shows alpha secondary structure, while in an AD affected amyloid protein it shows same tendency to become alpha and beta secondary structure. It means that for an AD affected amyloid protein, there is no definite secondary structure. Our work is a simulation-based approach to design a model for the detection of some diseases that occurs due to the deformation of protein structure. During

Table 7.3 Success rate with different training functions

Training functions	Success rate (%)	Average MSE attained in 1000 epochs
TRAINGDM	92	0.08999
TRAINGDX	94	9.03826×10^{-5}
TRAINGDA	98	9.71243×10^{-5}
TRAINLM	100	9.0084236×10^{-5}

the extraction and training of data, some difficulty may arise which can be rectified by using a properly configured ANN classifier. The ANN classifier with TRAINLM function shows the highest success rate as shown in Table 7.3.

7.6 Conclusion

In today's world, AD is a very common disease seen in elderly people. Proper detection of AD is necessary because till now cause and progression of AD is not well defined. Once it is defined in a proper way, drugs for this can be prepared. This proposed work can be taken as the framework for drug analysis and synthesis for diseases which arises due to the deformation of protein structure.

References

1. Berchtold NC, Cotman CW (1998) Evolution in the conceptualization of dementia and Alzheimer's disease, Greco-roman period to the 1960s. Neurobiol Aging 19(3):173–189
2. Davies L, Wolska B, Hilbich C, Multhaup G, Martins R, Simms G, Beyreuther K, Masters CL (1988) A4 amyloid protein deposition and the diagnosis of Alzheimer's disease: prevalence in aged brains determined by immunocytochemistry compared with conventional neuropathologic techniques. Neurology 38(11):1688–1693
3. Perneczky R, Guo LH, Kagerbauer SM, Werle L, Kurz A, Martin J, Alexopoulos P (2013) Soluble amyloid precursor protein β as blood-based biomarker of Alzheimer's disease. Transl Psychiatry 3(2):e227
4. Doecke JD, Laws SM, Faux NG, Wilson W, Burnham SC, Lam CP, Mondal A, Bedo J, Bush AI, Brown B, Ruyck KD, Ellis KA, Fowler C, Gupta VB, Head R, Caulay SL, Pertile K, Rowe CC, Rembach A, Rodrigues M, Rumble R, Szoeke C, Taddei K, Taddei T, Trounson B, Ames D, Masters CL, Martins RN (2012) Blood-based protein biomarkers for diagnosis of Alzheimer disease. Alzheimer's Dis Neuroimaging Initiative Austr Imaging Biomark Lifestyle Res Group, Arch Neurol 69(10):1318–1325
5. Barber RC (2010) Biomarkers for early detection of Alzheimer disease JAOA. J Am Osteopath Assoc 110:S10–S15
6. Caputa CB, Salama AI (1989) The amyloid proteins of Alzheimer's disease as potential targets for drug therapy. Microbiol Aging 10:451–461. http://www.dmso.org/articles/alzheimers/alzheim1.htm
7. Haykin S (2003) Neural networks A comprehensive foundation, 2nd edn. Pearson Education, New Delhi

Chapter 8
Identification of Stages of Industrial Sickness of Large- and Medium-Scale Units Using Certain Soft-Computational Approach

Deepak Goswami, Kandarpa Kumar Sarma and Padma Lochan Hazarika

Abstract Very often, industrial sickness is identified using certain traditional techniques which rely upon a range of manual monitoring and compilation of financial records. It makes the process tedious, time consuming, and often are susceptible to manipulation. Hence, decision makers, planners, and funding agencies of such units are sometimes surrounded by uncertainty and unpredictable situations while taking decisions regarding the state of industrial health and the subsequent measures required. Therefore, certain readily available tools are required which can deal with such uncertain situations arising out of industrial sickness. It is more significant for a country like India where the fruits of developments are rarely equally distributed. In this paper, we propose an approach based on certain soft-computational tools specially using Artificial neural network (ANN) to deal with industrial sickness with specific focus on a few such units taken from a less-developed northeast (NE) Indian state like Assam. More specifically, we here propose, a soft-computational tool which formulates certain decision support mechanism to decide upon industrial sickness using eight different parameters which are directly related to the stages of sickness of such units. The mechanism primarily identifies a few stages of industrial health using various inputs provided in terms of the eight identified parameters. This decision is further compared with the results obtained from another set of ANNs where the model uses certain signals and symptoms of industrial health to decide upon the state of a unit. Specifically, we train multiple ANN blocks with data obtained from a few selected units of Assam so that required decisions related to industrial health could be taken. The system thus formulated could become an important part of

D. Goswami (✉)
Department of Economics, Lalit Chandra Bharali College,
Guwahati 781011, Assam, India
e-mail: deepakgoswami321@gmail.com

K.K. Sarma
Department of Electronics and Communication Technology,
Gauhati University, Guwahati 781014, Assam, India
e-mail: kandarpaks@gmail.com

P.L. Hazarika
Department of Commerce, Gauhati University,
Guwahati 781014, Assam, India
e-mail: plhazarikagu@yahoo.co.in

K.K. Sarma et al. (eds.), *Recent Trends in Intelligent and Emerging Systems*, Signals and Communication Technology, DOI 10.1007/978-81-322-2407-5_8

planning and development. It can also contribute toward computerization of decision support systems related to industrial health and help in better management.

Keywords Industrial sickness · Classification · Artificial neural network (ANN) · Multilayer perceptron (MLP) · Mean square error (MSE)

8.1 Introduction

Industrial sickness is one of the key areas in Indian economy which have always threatened to lower the benefits of the economic growth currently observed. Specifically, in case of government-funded industrial sectors, the proper identification of sick industrial units play a critical role in planning and financial allocations [1–10]. Very often, industrial sickness is identified using certain traditional techniques which rely upon a range of manual monitoring and compilation of financial records. It makes the process tedious, time consuming, and often are susceptible to manipulation. Hence, decision makers, planners and funding agencies of such units are sometimes surrounded by uncertainty and unpredictable situations while taking decisions regarding the state of industrial health and the subsequent measures required. Therefore, certain readily available tools are required which can deal with such uncertain situations arising out of industrial sickness. It is more significant for a country like India where the fruits of developments are rarely equally distributed. In this paper, we propose an approach based on certain soft-computational tools specially using artificial neural network (ANN) to deal with industrial sickness with specific focus on a few such units taken from a less-developed northeast (NE) Indian state like Assam. More specifically, we here propose, a soft-computational tool which formulates certain decision support mechanism to decide upon industrial sickness using eight different parameters which are directly related to the stages of sickness of such units. The mechanism primarily identifies a few stages of industrial health using various inputs provided in terms of the eight identified parameters. This decision is further compared with the results obtained from another set of ANNs where the model uses certain signals and symptoms of industrial health to decide upon the state of a unit. Specifically, we train multiple ANN blocks with data obtained from a few selected units of Assam so that required decisions related to industrial health could be taken. Multiple ANN blocks are given the symptoms as inputs in codified forms with corresponding industrial health status as reference. A nonlinear mapping between these two domains are performed by the ANNs. These blocks, in the trained state, are able to differentiate between the industrial health stages as response to the inputs accumulated for three to four years for certain number of units. From these data, the ANN blocks are also able to provide a prediction of the likely state in the subsequent years.

The ANN is a nonparametric tool which can learn from the surrounding environment and use the knowledge thus acquired for subsequent decision making similar to biological systems [1]. This ability of the ANN has made it a popular mechanism for dealing with a range of applications in engineering, technology, finance, management, and related disciplines. The system thus formulated could become an important part of planning and development. It can also contribute toward computerization of decision support systems related to industrial health and help in better management.

The rest of the paper is organized as follows: Sect. 8.2 included the brief theoretical background relevant for the work. Primarily, this section focuses on different stages of industrial health (Sect. 8.2.1) and the basics of ANN (Sect. 8.2.2). Section 8.3 describes the details of the proposed soft-computational framework for analyzing industrial health. The experiments performed and the results derived are included in Sect. 8.4. Section 8.5 concludes the description.

8.2 Background

Here, we briefly discuss certain background notions relevant for the work. We focus on the definition and types of industrial health in Sect. 8.2.1. The basic concepts of ANN is presented in Sect. 8.2.2. The application of the multilayer perceptron (MLP), a feedforward ANN used for this work is covered in Sect. 8.2.2.1. Certain relevant details of the MLP and its training are covered in Sect. 8.2.2.2 which formulates the basic configuration of the system to configure the proposed soft-computational framework.

8.2.1 Industrial Sickness and Its Types

All the developing countries have assigned a very significant role to industrialization in their programmes of economic development. The reason is that the industrial sector is recognized as an indispensable tool of a stable economy. India has been trying to accelerate the rate of economic development through industrialization ever since the beginning of the planning era. A large number of industries have been setup during post-independence period. Despite governmental efforts in the form of various incentives, the rate of growth has failed to pick up [1]. Certain industries have been plagued by sickness corroding the narrow industrial base and is threatening the fragile industrial structure. This is more pertinent for the NE Indian state of Assam. Therefore, it is important to identify the stages of industrial sickness and clearly indicate its signals and symptoms. The different stages of sickness along with the determinants which identify these stages as per the guidelines of Reserve Bank of India (RBI) are given below [2]:

Table 8.1 Stages of industrial sickness and associated symptoms

Sickness	Production	Profit	Working	Loss	Marketing	Personnel	Net worth	Net worth
States	x^1	x^2	Capital, x^3	x^4	x^5	x^6	Rise, x^7	Fall, x^8
Normal (N)	↑	↑	↑	Nil	↑	↑	↑	Nil
Tending to sickness (TS)	↑	↓	NC	↑	NC	NC	NC	↑
Incipient sickness (IS)	↓	↓	↓	↑	↓	↓	↓	↑
Close (C)	↓	↓	↓	↑	Nil	Nil	Nil	↑

1. **Normal Unit** A normal unit is characterized by the efficient functioning of its functional areas like production, marketing, finance and personnel. In other words, a unit can be called healthy or in a normal state (NS) when it is earning profits, the current ratio is more than one, net worth is positive, and debt-equity ratio is good.
2. **Tending Towards Sickness** At this stage, a unit shows certain initial aberration in any of its functional areas. In other words, the unit faces some environmental constraints. At this time, the unit is said to be tending toward sickness (TS). The distinctive features of this stage are decline in profit in the last year as compared to the previous year and loss estimated in the current year.
3. **Incipient Sickness** The continuation of the deterioration in the functional areas of the unit, results in the actual setting in of industrial sickness. This stage is termed as incipient sickness (IS). At this stage, the unit incurs cash losses but imbalance in the financial structure may not be apparent.

These industrial health stages are summarized in the Table 8.1. The ↑ sign indicates growth, ↓ means fall, NC indicates no change, and Nil denotes zero value for particular sample industrial data.

8.2.2 ANN

Accuracy of a prediction and estimation process depends to a large extent on a higher value of proper classification. ANNs are one of the preferred classifiers for their ability to provide optimal solution to a wide class of arbitrary classification problems [11]. Multilayered perceptron (MLP) is a class of feedforward (FF) ANN trained with (error) backpropagation (BP). It has found widespread acceptance in several classification, recognition, and estimation applications [11].

8.2.2.1 Multilayered Perceptron Based Learning

The fundamental unit of the ANN is the McCulloch-Pitts neuron (1943). The MLP is the product of several researchers: Frank Rosenblatt (1958), H.D. Block (1962) and M.L. Minsky with S.A. Papart (1988). BP, the training algorithm, was discovered independently by several researchers (Rumelhart et al. (1986) and also McClelland and Rumelhart (1988)) [11].

A simple perceptron is a single McCulloch-Pitts neuron trained by the perceptron algorithm is given as:

$$O_x = g([w].[x] + b) \tag{8.1}$$

where $[x]$ is the input vector, $[w]$ is the associated weight vector, b is a bias value, and $g(x)$ is the activation function. Such a setup, namely the perceptron is able to classify only linearly separable data. A MLP, in contrast, consists of several layers of neurons held together by interlinked connectionist weights which are assigned values randomly at the beginning and updated continuously during training. These together generate a nonlinear computing. The connectionist weights retain the learning during training and facilitate an adaptation to reach the desired goal. The expression for output in a MLP with one hidden layer is given as:

$$O_x = \sum_{i=1}^{N} \beta_i g[w]_i.[x] + b_i \tag{8.2}$$

where β_i is the weight value between the ith hidden neuron. Such a setup maybe depicted as in Fig. 8.1. The process of adjusting the weights and biases of a perceptron or MLP is known as *training*. The perceptron algorithm (for training simple perceptrons) consists of comparing the output of the perceptron with an associated target value. The most common training algorithm for MLPs is BP which is an extended form of the method used to train the perceptron. This algorithm entails a backward flow of the error corrections through each neuron in the network.

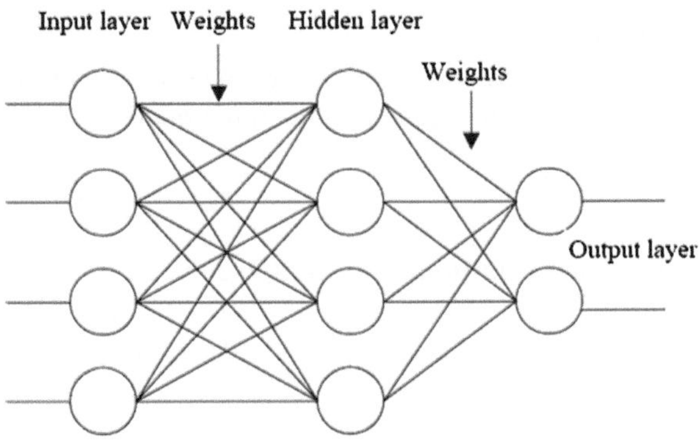

Fig. 8.1 Multilayer perceptron

In the most basic form, an ANN requires an input and should be associated with a reference which it tracks during each training iteration. The output, at the end of each training epoch, is compared with the reference and a corresponding adaptation of the connectionist weights are carried out. This is repeated till the output of the ANN is nearly equal to that of the reference [11].

8.2.2.2 Application of Error Back Propagation for MLP Training

The MLP is trained using BP algorithm depending upon which the connecting weights between the layers are updated. This adaptive updating of the MLP is continued till the performance goal is met. Training the MLP is done in two broad passes: one a forward pass and the other a backward calculation with error determination and connecting weight updating in between. Batch training method is adopted as it accelerates the speed of training and the rate of convergence of the mean square error (MSE) to the desired value [1]. The steps are as below:

- **Initialization**: Initialize weight matrix **W** with random values between $[-1, 1]$ if a tan-sigmoid function is used as an activation function and between [0, 1] if log-sigmoid function is used as activation function. **W** is a matrix of $C \times P$ where P is the length of the feature vector used for each of the C classes.
- **Presentation of training samples**: Input is $p_m = [p_{m1}, p_{m2}, \ldots, p_{mL}]$. The desired output is $d_m = [d_{m1}, d_{m2}, \ldots, d_{mL}]$.
  - Compute the values of the hidden nodes as:

$$\text{net}_{mj}^{h} = \sum_{i=1}^{L} w_{ji}^{h} p^{mi} + \text{Ø}_{j}^{h} \tag{8.3}$$

  - Calculate the output from the hidden layer as

$$o_{mj}^{h} = f_{j}^{h}(\text{net}_{mj}^{h}) \tag{8.4}$$

    where f(x)= $\frac{1}{e^x}$
    or f(x)= $\frac{e^x - e^{-x}}{e^x + e^{-x}}$
    depending upon the choice of the activation function.
  - Calculate the values of the output node as:

$$o_{mk}^{o} = f_{k}^{o}(\text{net}_{mj}^{o}) \tag{8.5}$$

- **Forward Computation**: Compute the errors:

$$e_{jn} = d_{jn} - o_{jn} \tag{8.6}$$

Calculate the MSE as:

$$\mathrm{MSE} = \frac{\sum_{j=1}^{M} \sum_{n=1}^{L} e_{jn}^{2}}{2M} \tag{8.7}$$

Error terms for the output layer is:

$$\delta_{mk}^{o} = o_{mk}^{o}(1 - o_{mk}^{o})e_{mn} \tag{8.8}$$

Error terms for the hidden layer:

$$\delta_{mk}^{h} = o_{mk}^{h}(1 - o_{mk}^{h})\sum_{j} \delta_{mj}^{o} w_{jk}^{o} \tag{8.9}$$

- **Weight Update**:
  - Between the output and hidden layers

$$w_{kj}^{o}(t+1) = w_{kj}^{o}(t) + \eta \delta_{mk}^{o} o_{mj} \tag{8.10}$$

where η is the learning rate ($0 < \eta < 1$). For faster convergence a momentum term (α) maybe added as:

$$w_{kj}^{o}(t+1) = w_{kj}^{o}(t) + \eta \delta_{mk}^{o} o_{mj} + \alpha(w_{kj}^{o}(t+1) - w_{kj}) \tag{8.11}$$

  - Between the hidden layer and input layer:

$$w_{ji}^{h}(t+1) = w_{ji}^{h}(t) + \eta \delta_{mj}^{h} p_i \tag{8.12}$$

A momentum term maybe added as:

$$w_{ji}^{h}(t+1) = w_{ji}^{h}(t) + \eta \delta_{mj}^{h} p_i + \alpha(w_{ji}^{o}(t+1) - w_{ji} \tag{8.13}$$

One cycle through the complete training set forms one epoch. The above is repeated till MSE meets the performance criteria. While repeating the above, the number of epoch elapsed is counted. A few methods used for MLP training includes gradient descent (GDBP), gradient descent with momentum BP (GDMBP), gradient descent with adaptive learning rate BP (GDALRBP), gradient descent with adaptive learning rate and momentum BP (GDALMBP), and gradient descent with Levenberg-Marquardt BP (GDLMBP).

8.3 Proposed Soft-Computational Framework for Identification of Stages of Industrial Sickness

We propose a soft-computational approach based on ANN to deal with industrial sickness so as to formulate certain decision support mechanism using eight different parameters. These parameters are directly related to the stages of sickness of such units. Specific focus is given to few such units taken from a less-developed NE Indian state like Assam. The mechanism is designed primarily to identify a few stages of industrial health using various inputs provided in terms of the eight identified parameters. This decision is further compared with the results obtained from field study and another approach where a different soft-computational model uses certain signals and symptoms of industrial health to decide upon the state of a unit. It primarily receives symptoms of industrial health of multiple units for three consecutive years. From such units, it generates the likely industrial health for the subsequent years. This mapping is performed during training which the ANN blocks retain for subsequent application.

The eight identified parameters are production volume (x^1), profit margin (x^2), working capital (x^3), loss margin (x^4), marketing (x^5), personnel (x^6), net worth rise (x^7), and net worth fall (x^8) as summarized in Table 8.1. The identified stages of an industrial unit are normal (N), tending to sickness (TS), incipient sickness (IS), and close (C). The association between the symptoms and the industrial health stages are distinctly shown by the Table 8.1. Figure 8.2 shows the composite block diagram of the proposed system. It has two parts: The first part involves the determination of the industrial health using eight different parameters. Figure 8.3 shows the system model

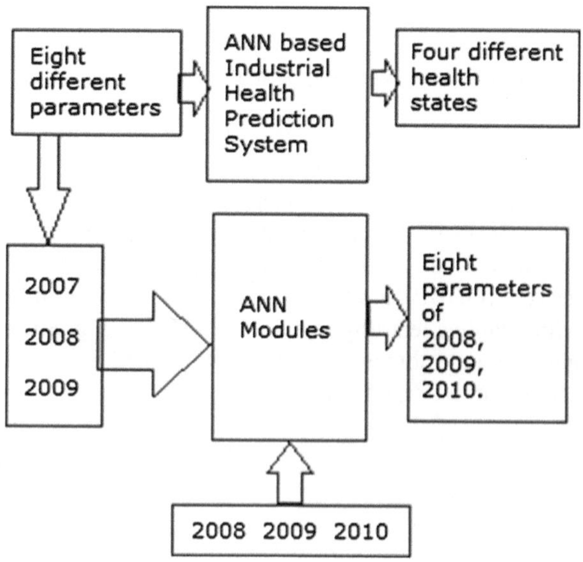

Fig. 8.2 Composite process logic of the complete work

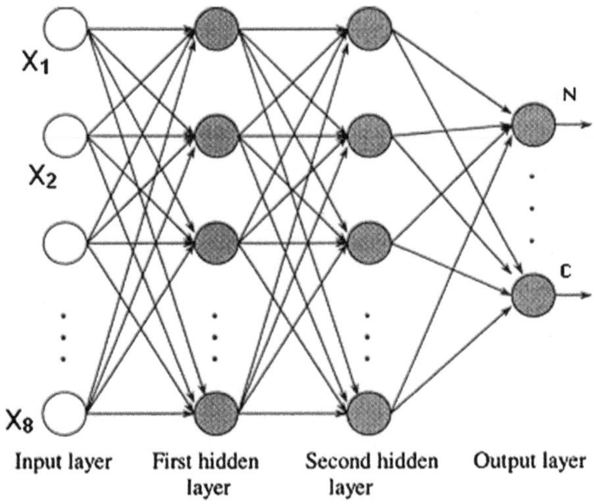

Fig. 8.3 Soft-computational framework to identify industrial health

Table 8.2 Specifications of the ANN-based model

Case	Item	Description
1	ANN type	Multilayer perceptron
2	Data size	Around 300 sets of data of 8×4 size each
3	Input vector size	16
4	Class	4
5	Number of hidden layers	2
6	Learning rate	0.1–0.5
7	Activation function	*Tansig-Logsig-Purelin*
8	MSE convergence goal	10^{-4}

used to formulate a decision support mechanism to identify the stages of sickness of industrial units. The specification of the MLP used to formulate the proposed industrial health identification system is summarized in Table 8.2.

Now, a set of five samples of 3 years of five numbers of industrial TS units are applied to another set of ANN blocks shown as in Fig. 8.4. Three ANN blocks, each for the years 2007, 2008 and 2009, receive inputs of five TS units and get references of 2008, 2009, 2010 respectively. The blocks are trained to certain number of sessions as required by specific goals fixed. During training, the ANN blocks capture the variations shown by the sample data presented to them. This is repeated for IS and *N* units as well. Table 8.3 shows certain training results related to the proposed approach. Each network is trained using GDLMBP as it proves to be computationally efficient. The blocks are designed to predict industrial health in the next year using the data available for a given year.

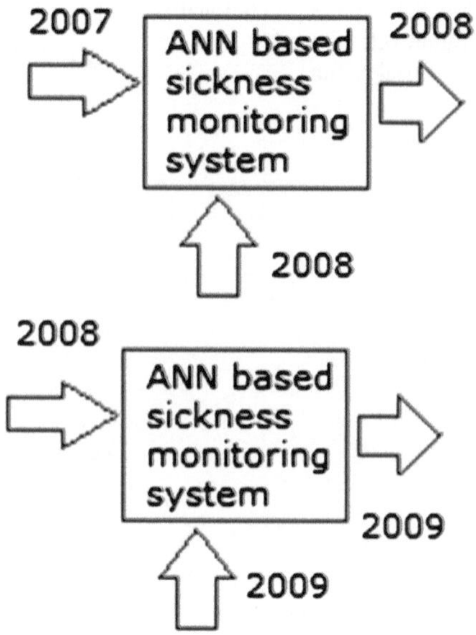

Fig. 8.4 Block diagram of an ANN framework to predict approaching industrial health

Table 8.3 Selection of appropriate training method and training results

SL Num	Epochs	MSE	Time (s)	Remark
GDBP	100	0.34	3.53	Goal not reached
	200	0.085	5.35	
	500	0.002	9.14	
	1000	0.001	17.3	
GDMBP	100	0.32	3.31	
	200	0.081	4.87	Goal not reached
	500	0.002	9.01	
	1000	0.001	16.4	
GDLMBP	10	0.0003	2.3	Goal reached after 10 (av.) epochs and within 3 s

8.4 Experimental Results and Discussion

The success of the proposed system is dependent on the configuration, training, data set and subsequent validation of the system. The configuration of the MLP formulating the system is summarized in Table 8.2. There are two hidden layers used in the system. The size of the first hidden layer is fixed with respect to number of neurons in the input layer and by following the considerations summarized in Table 8.4. It shows the performance obtained during training by varying the size of the hidden layer.

Table 8.4 Performance variation after 1000 epochs during training of an ANN with variation of size of the hidden layer

Case	Size of hidden layer (× input layer)	MSE attained	Precision attained in %
1	0.75	1.2×10^{-3}	87.1
2	1.0	0.56×10^{-3}	87.8
3	1.25	0.8×10^{-4}	87.1
4	1.5	0.3×10^{-4}	90.1
5	1.75	0.6×10^{-4}	89.2
6	2	0.7×10^{-4}	89.8

The case where the size of the hidden layer taken to be 1.5 times to that of the input layer is found to be computationally efficient. Hence, the size of the hidden layer is fixed at 1.5 times to that of the input layer. The next layer also is provided a length of 1.5 times to that of the first hidden layer which is fixed following the considerations as summarized in Table 8.4.

Table 8.3 shows certain training results related to the proposed approach. For training the ANN as part of the industrial health identification system, the conditions and contents of the Table 8.1 are codified using binary logic. For example, the growth state is represented by 11, fall by 10, NC denoted by 01 and Nil 00. The parameters have a set of codes as described while the industrial health have another group of unique class states. The process logic is described by the block diagram shown in Fig. 8.3.

The proposed approach is a decision support system, hence it is likely to face multiple situations which it must learn first and be ready for application. Hence, speed and low computational complexity is critical. It is directly linked to proper selection of the training algorithm which is of utmost importance. The values in Table table3 indicate that the GDMBP is marginally faster than GDBP. The fastest training is carried out by GDLMBP. It achieves the MSE convergence goal within the first ten epochs. Figure 8.5 shows training time performance of the system using GDLMBP. Similarly, Fig. 8.6 shows gradient descent, Marquardt adjustment parameter (μ), and validation checks during training. As the GDLMBP turns out to be the fastest and the most reliable training method, the ANN training using it is used for subsequent testing. The ANN-based training framework shows a success rate of around 95–98 % precision. The trained ANN is tested with a set of data as shown in Tables 8.5, 8.6 and 8.7. It represents average sets of data of stages showing small-, medium-, and large-scale units tending toward sickness over a 3-year period. The data clearly show that while working capital remains constant, profit starts to fall. It is a sign of TS. The ANN provides accurate prediction of industrial health for all the 30 sets of training data considered for validation of performance. At the end of atleast 15 trials, the success rate of around 95–98 % is consistently achieved. Some of the related works have been reported in [12].

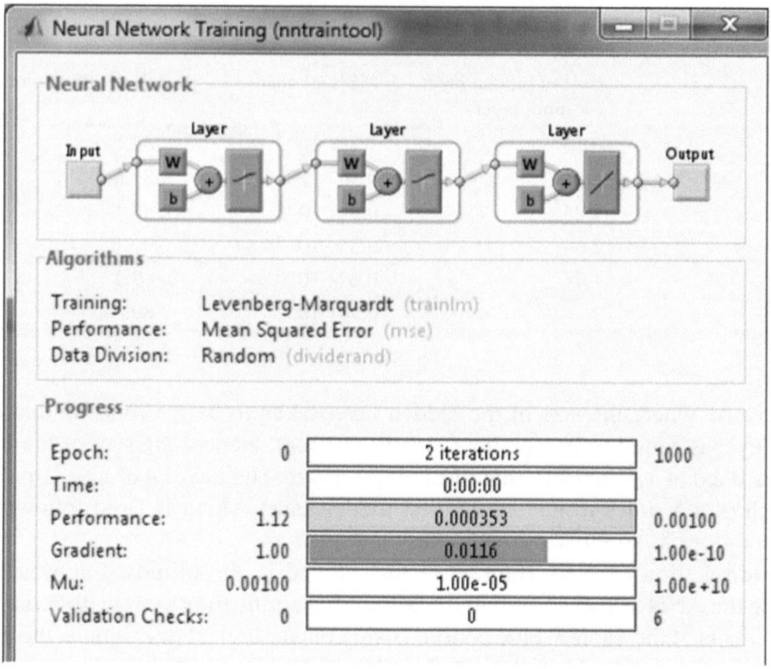

Fig. 8.5 Training performance using GDLMBP

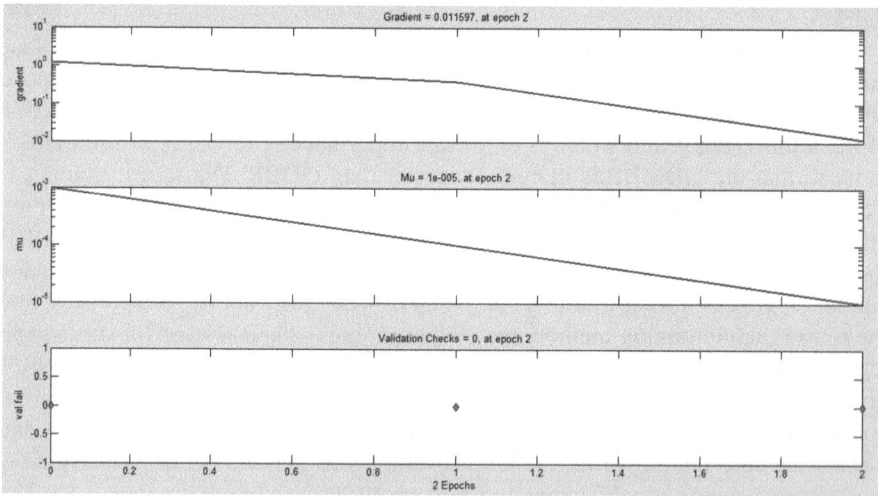

Fig. 8.6 Training validation and other performance curves

Table 8.5 Testing dataset used to validate the training of the ANN-based system

Sickness states	Production (tons)	Profit (%)	Working capital Rs. $\times 10^6$	Loss (%)	Marketing	Personnel	Net worth rise (%)	Net worth fall (%)
2007	50	6	7	0	3	15	1.5	0
2008	80	3.7	7	0	3	15	0.5	0
2009	85	0.3	6	0	2	12	0.2	0
2010	81	0.0	2	3	0	8	0	1.5

The data reflect average values of five small-scale units tending toward sickness

Table 8.6 Testing dataset used to validate the training of the ANN-based system

Sickness states	Production (tons)	Profit (%)	Working capital Rs. $\times 10^6$	Loss (%)	Marketing	Personnel	Net worth rise (%)	Net worth fall (%)
2007	140	6.1	43	0	11	62	1.3	0
2008	135	4.3	41	0	10	61	1.3	0
2009	128	0.8	36	0	7	52	0.8	0
2010	101	0.0	18	2.2	0	38	0	1.7

The data reflect average values of five medium-scale units tending toward sickness

Table 8.7 Testing dataset reflecting average values of five large-scale units tending toward sickness

Sickness states	Production (tons)	Profit (%)	Working capital Rs. $\times 10^6$	Loss (%)	Marketing	Personnel	Net worth rise (%)	Net worth fall (%)
2007	241	6.3	305	0	20	165	1.1	0
2008	231	5.5	289	0	18	158	1.0	0
2009	203	1.3	275	0	13	155	0.7	0
2010	191	0.0	238	2.1	5	141	0	1.4

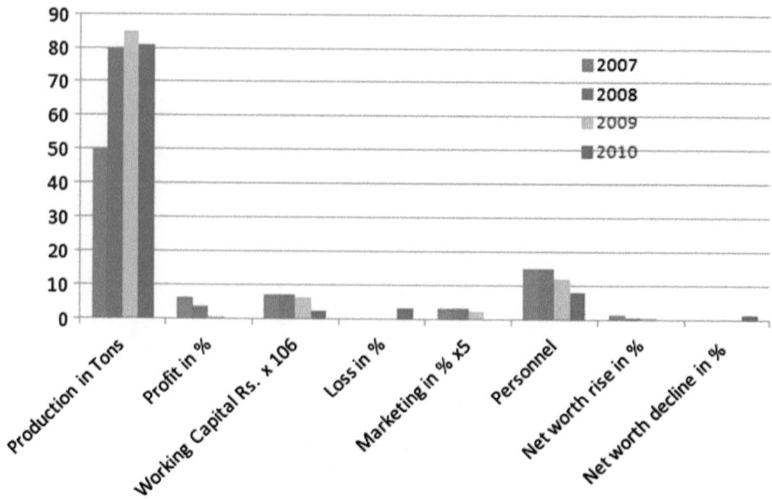

Fig. 8.7 Training validation and other performance curves

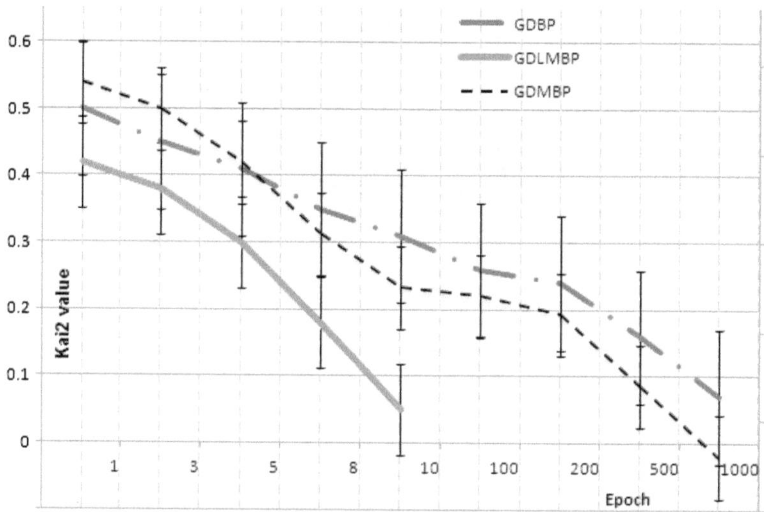

Fig. 8.8 χ^2 plot between actual and desired results for three different training methods

The second stage of the system is formed by an ANN framework to predict approaching industrial health as described in Fig. 8.4.

A set of results are derived using the system to predict industrial health in the immediate future. When the trained set of ANNs are tested (Sect. 8.3) and summarized in the block diagram shown using the data samples in Table 8.5, the predicted results generated closely match the expected results. It shows an accuracy of around 95 % range. A set of these values are shown in Fig. 8.7. A χ^2 plot between actual and desired results for three different training methods is shown in Fig. 8.8. The plot also includes error bars for each training method. For GDMLBP, the error falls to satisfactory level within the first ten epochs which reflect the efficiency of the system in terms of computational cost and accuracy. The proposed system thus not only identifies the industrial health of sample units considered but is also able to predict the approaching state in the immediately following year. The effectiveness of the system is ascertained by the accuracy it shows in identifying the sickness stages and the industrial health likely to be achieved in the near future. The above results have been derived using field survey data of around 315 medium- and small-scale units taking 10 % random samples for training, 15 % for validating, and 20 % for testing. Thus, the framework proposed is effective for medium- and small-scale units in identifying the industrial health and the approaching state in the immediate future.

8.5 Conclusion

Here, we have proposed a soft-computational framework for identifying the present industrial health of production units and the approaching state. The proposed framework is formulated with specific focus on a few industrial units taken from a less-developed NE Indian state like Assam. More specifically, the proposed soft-computational framework formulates certain decision support mechanism to decide upon industrial sickness using eight different parameters which are directly related to the stages of sickness of such units. The mechanism primarily identifies a few stages of industrial health using various selected inputs which are critical in effective functioning of any related decision support system. The proposed framework, next, can be extended to include a diverse range of industrial states covering micro- and large-scale units and modified enough to provide forecasts of more number of immediately following years.

References

1. Economic Survey, Assam, 2009–10. Directorate of Economics and Statistics, Govt of Assam, Guwahati, Assam, India, p 93
2. Nayak SS, Misra RN (2006) Sickness in small-scale industrial units and its revival. Orissa Review
3. Junejo MA, Rohra CL, Maitlo GM (2007) Sickness in small-scale industries of Sindh: causes and remedies, a case study of Larkana estate area. Aust J Basic Appl Sci 1(4):860–865

4. Parmar RS (1995) Industrial sickness in small scale sector, Dattsons Publishers, Naqpur
5. Panda RK, Meher R (1992) Industrial sickness, a study of small-scale industries. APH Publishing, New Delhi
6. Kadam JJ, Laturkar VN (2011) A study of financial management in small scale industries in India. Int J Exclus Manag Res 1(3)
7. Singh N (2011) Industrial sickness: causes and remedies. Ann Manag Res 1(2)
8. Dhalokia BH (1989) Industrial sickness in India: weed for comprehensive identification criteria. Vikalpa 14:2
9. Jagtap KN (2011) Impact of globalization on small scale sector. Indian Streams Res J I(IX)
10. Mishra RN, Nayak SS (2008) Industrial sickness in small scale sector. Discovery Publishing House Pvt Ltd, New Delhi
11. Haykin S (2006) Neural Networks, 2nd edn. Pearson Education, New Delhi, India
12. Goswami D, Hazarika PL, Sarma KK (2012) Stages and symptoms of industrial sickness-a mathematical model applied to a few small scale industrial units in ne Indian state of Assam. In: Proceedings of 13th WSEAS international conference on mathematics and computers in business and economics, Iasi, Romania, pp 152–157

Part III
HCI and Bio-inspired System Design

This section is formed by works of Mazumdar et al. and Sarma et al. In the work, Multi-core Parallel Computing and DSP Processor for the Design of Bio-inspired Soft Computing Framework for Speech and Image Processing Applications, the authors have highlighted how the learning process of ANN is dependent on hardware and data volume. Similarly, the chapter by Mazumdar et al. describes an important design for smooth human–computer interaction (HCI) applications.

Chapter 9
Adaptive Hand Segmentation and Tracking for Application in Continuous Hand Gesture Recognition

Dharani Mazumdar, Madhurjya Kumar Nayak and Anjan Kumar Talukdar

Abstract Hand gesture recognition system is an essential element used for human–computer interaction (HCI). The use of hand gestures provides an attractive alternative to cumbersome interface devices for HCI. Proper hand segmentation from the background and other body parts of a video stream is the primary requirement of the design of a hand gesture-based application. These video frames are captured from a low-cost webcam (camera) for using in a vision-based gesture recognition system. This work reports the design of a continuous hand gesture recognition system. The paper also includes the description of a robust and efficient hand segmentation algorithm where a new method for hand segmentation using different color space models as required by morphological processing are utilized. Problems such as skin color detection, complex background removal, and variable lighting conditions are found to be efficiently handled by this system. Noise present in the segmented image due to dynamic background can be removed with the help of this adaptive technique. The proposed approach is found to be effective for a range of conditions.

Keywords Human–computer interaction (HCI) · Adaptive hand segmentation · Hand gesture recognition · Color space models · Morphological processing

9.1 Introduction

Gestures are specific motions of body part/parts that represents some meaningful message. Gesture recognition is a mechanism through which a machine can understand the meaning of any gesture. Hand gesture can be subdivided into two types.

D. Mazumdar (✉) · M.K. Nayak · A.K. Talukdar
Department of Electronics and Communication Engineering,
Gauhati University, Guwahati 781014, Assam, India
e-mail: dharanimazumdar89@gmail.com

M.K. Nayak
e-mail: madhurjya.nayak@yahoo.com

A.K. Talukdar
e-mail: anjan.nov@gmail.com

K.K. Sarma et al. (eds.), *Recent Trends in Intelligent and Emerging Systems*,
Signals and Communication Technology, DOI 10.1007/978-81-322-2407-5_9

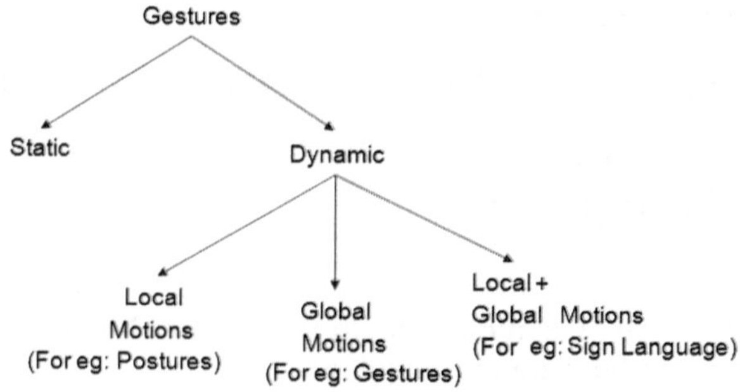

Fig. 9.1 Hand gesture types

First global motion where the entire hand moves; whereas in the second, one only local motion (or posture) is available where only the fingers move [1–3]. Figure 9.1 shows the various types of hand gestures.

Waving goodbye is a gesture. Pressing a key on a keyboard is not a gesture because the motion of a finger on its way to hitting a key is neither observed nor significant. All that matters is which key was pressed.

Today's world have many applications of gesture recognition systems such as human–computer interaction (HCI), robotic arm control, gaming consoles and television control mechanisms, sign language, etc. [4]. With such widespread applications, it is imperative for us to study and to make this system as user friendly as possible. Hand segmentation is one of the most important processes in a gesture recognition system because a properly segmented image of a region of interest (ROI), i.e., hand can provide better performance. It means that an effectively extracted segmented part of the image aids detection performance of a gesture recognition system.

Hand gesture recognition systems are important segments of HCI. The use of hand gestures provides an attractive alternative to cumbersome interface devices for HCI. Proper hand segmentation from the background and other body parts of a video stream is the primary requirement of the design of a hand gesture-based application. These video frames are captured from a low-cost webcam (camera) for use in a vision-based gesture recognition system. This work reports the design of a continuous hand gesture recognition system using a robust and efficient hand segmentation algorithm. Here a new method for hand segmentation using different color space models by required morphological processing is utilized. Problems such as skin color detection, complex background removal, and variable lighting condition are found to be efficiently handled with this system. Noise present in the segmented image due to dynamic background can be removed with the help of this adaptive technique. The proposed approach is found to be effective for a range of conditions.

The rest of the paper is organized as follows. Section 9.2 provides a brief review of the design aspects required for the work. Section 9.3 describes the proposed model. Section 9.4 contains the experimental results. Section 9.5 concludes the work.

9.2 Design Notions and Relevant Literature

A generic hand posture recognition system can be implemented using the block diagram as shown in Fig. 9.2.

- Image Acquisition-The image frame is taken from the input video feed and then processed for the system to get it as a gesture.
- Image Segmentation and Tracking- Image segmentation is one of the most important steps in this system as without proper segmentation; we cannot get a proper gesture input to the system and hand tracking is a high resolution technique that is employed to know the consecutive position of the hands of the user and hence represent objects in 3D.
- Feature Extraction-After the successful tracking, there is a need to extract the important feature points from the available data points of the track path. In pattern recognition and in image processing, feature extraction is a special form of dimensionality reduction.
- Classification-Classification plays a vital role in gesture recognition process. It is a statistical method that takes feature set as input and gives a class labeled output, which are the required output gestures.

Now there are various models which are followed to get the segmented output. Some of them are as follows:

1. Color Segmentation: In this model, we first take the input image and then convert it into HSI or YCbCr color space as color intensity in RGB has to be controlled individually but in YCbCr color space, y controls the intensity. Then the intensity is adjusted to match the color required (which in our case would be the color of our hand). Next, thresholding is done on the image and is converted into a binary image. Noise is minimized using morphological operations like erosion, dilation, etc. Disadvantage of this system is that if the background has any object having the same color as the hand, noise will be very high [5].
2. Background Subtraction: In this model, we first take the image of the background and store it. Now, when we get an image frame then the image is subtracted from the previously stored background. This gives only the moving or dynamic parts which in our case would be the body parts. Disadvantage of this system is that if the lighting conditions change abruptly then there is a change in pixel value

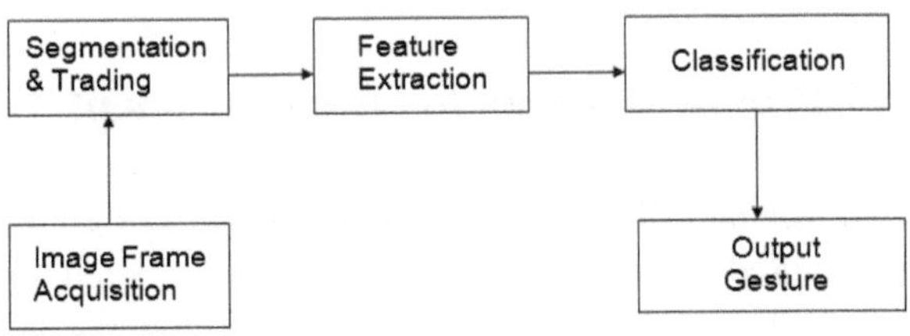

Fig. 9.2 Block diagram for gesture recognition system

where the light intensity is changed and additive noise contributes to the output [6, 7].

3. Object tracking with HSV color scheme:In this model, along with color-based segmentation, a hand tracking mechanism is used. Now through this mechanism, we first track the hand and then segment it out using color segmentation. This method is more advantageous over both the above mechanisms in tracking. But this model also has a disadvantage that when the motion of the hand which is ROI in our case moves too fast then the system fails to track it, thus add noise to the output of this process [8].

9.2.1 Previous Work Done

Some of the related works included here.

In [9], an HMM is employed to recognize the tracked gesture for controlling desktop applications like games, painting programs, and browsers. In [10], a scheme for isolated word recognition in the Japanese Sign Language (JSL) is reported. The authors apply HMMs to model the gesture data from right and left hand in a parallel mode. The information is merged by multiplying the resulting output probabilities. In [11], instead of using colored gloves, 3D object shape and motion extracted features with computer vision methods as well as a magnetic tracker fixed with the signer's wrists are used. They introduce a parallel algorithm using HMMs in order to model and recognize gestures from continuous input stream. Shape, location, and trajectory of left hand, in addition to location and trajectory of right hand are implemented using separate channels of HMMs. Each channel has been learned with relevant data and combined features. Moreover, individual networks of HMMs have been constructed for each channel and a modified Viterbi algorithm was employed to search within the networks in parallel. From each network, the trajectory probabilities with the same word sequences are combined together. In [12] a proposed system where a Kalman filter is used for hand tracking to obtain motion descriptors and hand region. This model is fairly robust to background cluster and uses skin color for hand gesture tracking and recognition. In HMMs, the current observations are based only on the current state, but the current observations for the Maximum Entropy Markov Models (MEMMs) that is proposed by McCallum et al. depend on the previous and the current states [13].

9.3 Proposed Model for Hand Segmentation

Here, we describe the proposed model for hand segmentation. At first, image frame is captured using a camera. Next, this image is fed to background subtraction block which minimizes the noise due to the background containing objects having the same color as that of the hand. In this algorithm, a colored glove is worn on the palm part

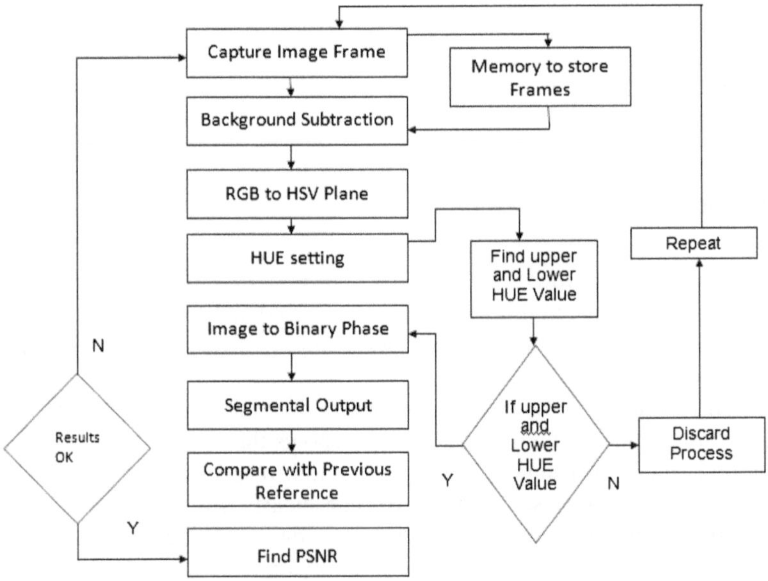

Fig. 9.3 Flow chart for proposed model

of the users hand. HSV color plane is the most suitable color plane for color-based image segmentation. So, we have converted the color space of original image frame of the camera, i.e., from RGB to HSV plane. Now setting the hue value ranges from the lower threshold to upper threshold for the particular color of hand gloves, we can easily eliminate all the other parts of the image frames. Finally, this output is converted to binary form and certain morphological operation likes erosion, dilation, etc., are carried out. It results in a noiseless segmented hand or ROI that can be used subsequently. Then, the centroid of the segmented hand is determined and subsequently the centroid is used for making gesture trajectories. Figure 9.3 shows flowchart of this algorithm. The related algorithm can be summarized as below:

Step 1. The input is first dynamically stored in an array. Then the frame is compared with the current frame. It gives us the moving parts only as output.
Step 2. This image is then converted to HSV color scheme.
Step 3. Hue value is then set properly for the color of the hand glove or the cap of the finger. If the range of hue value is sets properly go to next step, else repeat the process.
Step 4. HSV plane is converted to binary plane with proper thresholding.
Step 5. Certain morphological operations on the binary plane gives a proper segmented result.
Step 6. Compare the segmented result with references. If result is satisfactory, then calculate the PSNR value; otherwise return back to step 1.

The input will then contain the hand and some noise due to dynamic lighting conditions, if any, in the image. This thresholding uses an adaptive local filtering mechanism to minimize this noise by working on the image pixel by pixel.

9.4 Results and Discussions

A few samples are recorded by a webcam and indoor–outdoor sets are created for use with this algorithms. In this paper, we have included the segmented output result found from bare hand and gloved hand. The samples consider illumination and background variation.

The results obtained from the experiments are shown in Fig. 9.4. Figure 9.4a shows the input frame to the system. Figure 9.4b, c shows the outputs of color segmentation method and background subtraction method, respectively. Figure 9.4d shows the output from the proposed system. We can clearly see that noise is much less in the output of the proposed model. Here, we have considered dynamic lighting conditions. Again, Fig. 9.5 shows some output obtained during testing. To determine the robustness of the system, various environments are considered with and without glove. Figure 9.4a, b shows the input and output of the system respectively in without glove and Fig. 9.5c, d shows the input and output of the proposed model when, i.e., with glove. Table 9.1 shows the comparative output obtained from the proposed model. All the systems are tested under various conditions and the outputs are given in a qualitative manner.

Computational performance is summarized in Table 9.2. It also shows the comparative noise reduction performance of the proposed model.

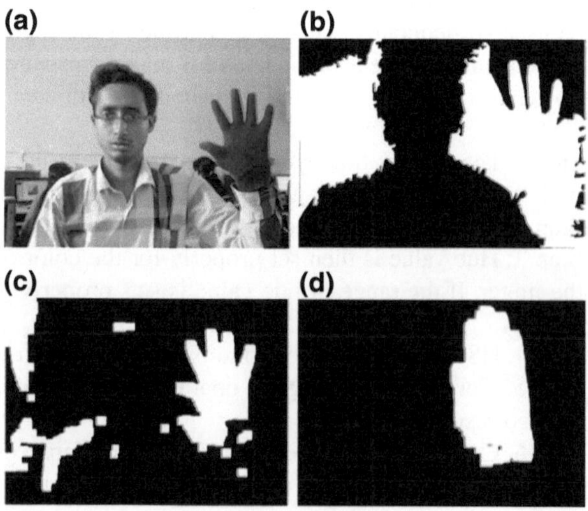

Fig. 9.4 Segmented output from proposed model. **a** Original image. **b** Color segmented output. **c** Output from backqround subtraction. **d** Output from proposed model

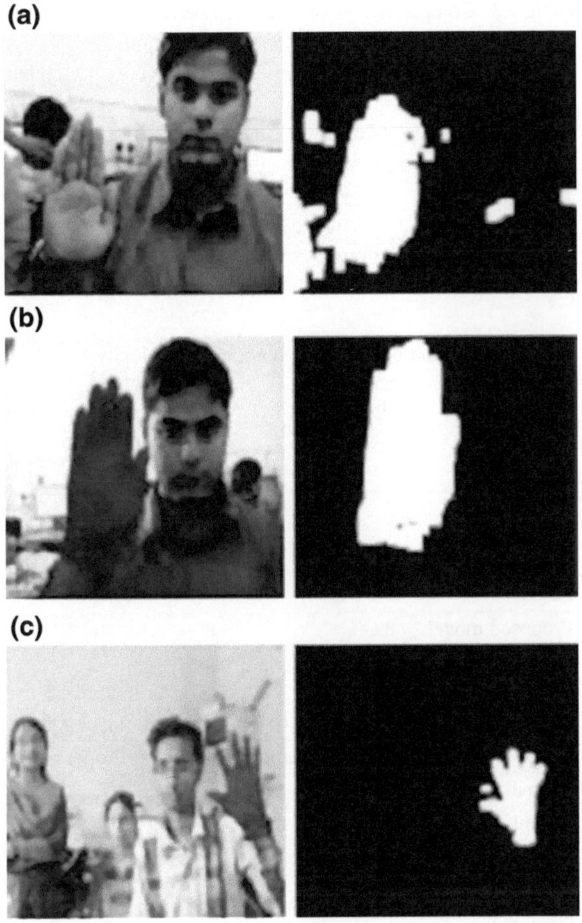

Fig. 9.5 Comparison of segmented output. **a** Result at normal condition (naked hand). **b** Result at normal condition (gloved hand). **c** Result at complex background

Table 9.1 Comparison among outputs obtained from different approaches

Models	Normal	Complex background	Dynamic lighting	Tracking fast hand movements
Color based	Working	Working	Working	Working
Background subtraction	Working	Working	Sensitive	Sensitive
Object tracking based	Working	Working	Robust	Sensitive
Proposed model	Working	Robust	More robust than earlier	Robust

Table 9.2 Comparative noise reduction performance of the proposed model

Models	Normal (PSNR)	Low light (PSNR)	Complex background (PSNR)	Dynamic lighting (PSNR)
Input (PSNR)	27.6362	27.6362	27.4347	27.6362
Color based	28.7693	30.2892	28.2881	28.0433
Background subtraction	27.9975	29.8822	30.2125	27.8314
Proposed model	31.0983	32.2328	31.4188	32.0660

Table 9.3 Comparison of computational complexity

Models	Normal (Delay in %)	Complex background (Delay in %)	Dynamic lighting (Delay in %)	Tracking fast hand movements (Delay in %)
Color based	2	10	10	8
Background subtraction	3	0	8	10
Object tracking based	4	12	8	8
Proposed model	8	8	8	8

Table 9.4 Statistical efficiency of our proposed model at different conditions

Condition	Test sample	Proper output	Efficiency (%)
Normal	10	10	100
Complex background	10	8	80
Dynamic lighting	10	9	90
Tracking fast hand movements	10	8	80

Complexity is one of the critical aspects of all systems or algorithms. It is the measure of how much work is required to solve different problems. Here, we have observed the required processing time for the proposed algorithm under different condition. We represent it as a function of delay as shown in Table 9.3. Excepting the condition for complex background, our proposed model shows no delay, i.e., it can work as a real-time process. In case of other techniques, for some conditions, they show some delay. Table 9.4 shows the efficiency of our proposed model under various conditions. We have tested the segmented result atleast 10–15 times for each condition. Among the given four conditions, complex background and tracking fast hand movement are the most crucial ones. With glove-based technique, we get the advantage that the gesture is not bound to wear full-sleeved clothes as the hue setting is adjusted for the color of the glove only. Earlier models such as color segmentation had problems when the background contained certain objects which had color almost similar to ROI. Also background subtraction failed to give suitable output when dynamic lighting conditions are taken into consideration. Also sometimes, the input gesture maybe fast. This makes detection through object-based tracking process different, because the patterns contain large amount of noise. The proposed algorithm can deal with these problems efficiently.

9.5 Conclusion

Here, we proposed an adaptive algorithm for hand segmentation and tracking of continuous hand posture recognition. We have tested the system under illumination and background variations. The results establish the effectiveness of the system. But like other known system of similar type, dynamic background puts restriction on the system performance. Subsequent work shall focus on this aspect.

References

1. Ishihara T, Otsu N (2004) Gesture recognition using auto-regressive coefficients of higher-order local auto-correlation features. In: Proceedings of the sixth IEEE international conference on automatic face and gesture recognition, pp 583–584
2. Florez F, Garcia JMI, Garcia J, Hernandez A (2002) Hand gesture recognition following the dynamics of a topology-preserving network. In: Proceedings of fifth IEEE international conference on automatic face and gesture recognition, Washington, pp 318–319
3. Lee HK, Kim JH (1999) An HMM-based threshold model approach for gesture recognition. IEEE Trans Pattern Anal Mach Intell 21(10):961–963
4. Yang HD, Sclaroff S, Lee SW (2009) Sign language spotting with a threshold model based on conditional random fields. IEEE Trans Pattern Anal Mach Intell 31(7):1264–1265
5. Phung SL, Bouzerdoum A, Chai D (2005) Skin segmentation using color pixel classification: analysis and comparison. IEEE Trans Pattern Anal Mach Intell 27(1):148–151
6. Manresa C, Varona J, Mas R, Perales FJ (2000) Real-time hand tracking and gesture recognition for human-computer interaction. In: Computer vision center Universitat Autonoma de Barcelona, Barcelona, Spain, pp 1–3
7. Lu YC (2011) Background substraction based segmentation using motion feedback. In: Proceedings of first international conference on robot vision and signal processing (RVSP), Kaohsiung, pp 224–227
8. Emami E, Fathy M (2011) Object tracking using improved CAMShift algorithm combined with motion segmentation. In: Proceedings of 7th Iranian machine vision and image processing (MVIP), Tehran, pp 1–4
9. Keskin C, Aran O, Akarun L (2005) Real time gestural interface for generic applications. In: Proceedings of European signal processing conference (EUSIPCO Demonstration Session), Antalya
10. Tanibata N, Shimada N, Shirai Y (2002) Extraction of hand features for recognition of sign language words. In: Proceedings of international conference on vision interface, pp 391–398
11. Vogler MD (2001) A framework for recognizing the simultaneous aspects of American sign language. J Comput Vision Image Underst 81(3):358–384

12. Binh ND, Shuichi E, Ejima T (2005) Real-time hand tracking and gesture recognition system. In: Proceedings of international conference on graphics, vision and image processing (GVIP-05), Egypt, pp 362–368
13. McCallum A, Freitag D, Pereira F (2000) Maximum entropy markov models for information extraction and segmentation. In: Proceedings of international conference on machine learning, pp 591–598

Chapter 10
Multicore Parallel Computing and DSP Processor for the Design of Bio-inspired Soft Computing Framework for Speech and Image Processing Applications

Dipjyoti Sarma and Kandarpa Kumar Sarma

Abstract Real-time applications like speech processing and image processing are known to have hardware dependencies. Again artificial neural network (ANN)-based recognition systems show dependance on data and hardware for achieving better performance. Therefore, there always exist the possibility of exploring means and methods to evaluate performance difference achieved by ANNs for speech- and image processing applications when hardware framework is varied and at times enhanced and expanded. Formation of certain parallel processing architectures based on multicore layouts are described here. With such frameworks, certain ANN-based processing of speech and image inputs are carried out. The results derived show that the capability of the ANN improves with large sizes of data and expanded hardware layouts. Similar approach is also followed using digital signal processor (DSP). The experimental study shows that compared to conventional CPUs, DSP processor architectures like TMS320C6713 provide greater processing speed and throughput.

Keywords Artificial neural network (ANN) · Digital signal processor · Parallel processing architectures · Speech · Image

10.1 Introduction

Application of digital signal processing and certain bio-inspired soft-computing tools such as artificial neural network (ANN) on speech and image signals demands high computing requirements. Computation of ANN resembles brain. As in brain, the

D. Sarma (✉)
Department of Electronics and Communications Engineering,
Don Bosco College of Engineering and Technology,
Assam Don Bosco University, Guwahati, India
e-mail: dipsarma4u@gmail.com

K.K. Sarma
Department of Electronics and Communication Technology,
Gauhati University, Guwahati 781014, Assam, India
e-mail: kandarpaks@gmail.com

K.K. Sarma et al. (eds.), *Recent Trends in Intelligent and Emerging Systems*,
Signals and Communication Technology, DOI 10.1007/978-81-322-2407-5_10

ANN also employs many computational elements that works concurrently, and finally achieves a distributed and parallel processing structure. ANN that performs speech recognition and synthesis, or pattern classification consist of large number of neurons and inputs. Every neuron computes a weighted sum of its inputs and applies a nonlinear function to its result [1]. The ANN recognizes patterns based on information and weights during training. However, the use of ANN classifier remains constrained due to the availability of powerful hardware to provide sufficient speed during training. The basic operation performed by a neuron during classification can be written as

$$Y = f\left(\sum_i x_i * w_i + b\right) \tag{10.1}$$

where x_i is the input, w_i is the connectionist weight, b is the bias, and $f(.)$ is the activation function.

Thus for each classification, the network must perform one multiplication and one addition for every connection which translates to a few billion multiply add operations per second. Only parallel implementations, in which several connections are evaluated concurrently, achieve such computational power [2]. General-purpose personal computers (GPPC) and workstations are the most popular computing platforms used by researchers to simulate ANN algorithms. They provide a convenient and flexible programming environment and technology advances have been rapidly increasing their performance and reducing their cost [3]. But ANN simulations for image and speech signals can still overwhelm the capabilities of even the most powerful GPPC.

There is always a need for computational (or robotic) systems to process information in real time and achieve their goals in a way that is efficient, self-optimizing, adaptive, and robust in the face of changing inputs. An ANN-based bio-inspired speech and image recognition system works in real time in a dynamic manner to recognize human beings using distinctive image and speech inputs. During the last few decades, several works have been reported on speech and image processing where ANN have been extensively used. The use of ANN leads to lower processing speed performance and remains constrained due to the availability of powerful hardware to provide sufficient speed during training. So some fast processing techniques for application of ANN and other related works for speech and image recognition are yet to be revealed deeper. In the paper [2], parallel processing of speech analysis techniques using networked computers is considered by the authors. The authors in survey paper [1], give a survey and suggest a method of implementations of ANN, utilizing the most recent and accessible parallel computer hardware and software. However, the features like speed, cost-effectiveness, reprogram ability, energy efficiency, etc., have made the DSP processors more suitable and advantageous for application in bio-inspired soft-computing tool design.

The implementation of ANN requires large processing time in training phase. This creates barrier in real-time implementation of such applications. The huge amount of data required for image processing and computer vision applications present a significant problem for conventional microprocessors. Similarly, speech processing also shows data and hardware dependence. Both speech and image signals require large storage capacity and processing for real-time applications and implementations. This dependence on storage and processing capability is more obvious in recognition systems based on learning. In the recognition phase, the implementation of ANN requires large training time when data size is bigger. The researchers have reported that learning of ANN improves with large data size but for generation of greater processing speed hardware becomes critical. Faster learning of ANN is dependent on the number of nodes used. A detailed literature survey shows that, for faster processing of ANN-based speech and image application certain specialized hardware like DSP processor or parallel computing environment-based framework can be designed. In this chapter, we certain experimental study on performance of frameworks using parallel computing environment and DSP processor that leads to the design of a bio-inspired system. We have examined whether the proposed approaches show improvement compared to a conventional framework in terms of computational capability and recognition rate. We specially focus on the implementation of certain ANN-based applications involving image and speech inputs. The proposed architectures show distinct advantage in terms of processing power and cognitive capability as compared to conventional approach of implementing a soft-computational framework like ANN for image and speech applications. In terms of computational capability, the approach using DSP processor shows around 8 % improvement compared to a four-core parallel computing framework. Similarly, improvements in recognition rate are around 4 % with applications involving speech and image samples. The rest of the chapter is organized as below.

Section 10.2 provides a brief theoretical background. Section 10.3 includes the results and discussion. The process flow of the work is also described here. The work is concluded in Sect. 10.4.

10.2 Theoretical Consideration

Here, we briefly describe about the basic theoretical notions related to the work. Section 10.2.1 discusses about ANN. In Sect. 10.2.2, a brief description of parallel processing is given. A brief description of DSP processor and their advantages is provided in Sect. 10.2.3

10.2.1 ANN

An ANN is a mathematical tool or computational model based on the analogy of biological nervous systems [4] consisting of interconnected group of artificial neurons and information processing using a connectionist approach. An ANN may be single or multilayered. The knowledge gained during the training phase is stored in the interconnecting neurons and used subsequently.

10.2.2 Parallel Processing

Many contemporary applications are driven by imagery [5] and speech inputs. A rapid increase in image sizes and processing requirements necessitates parallel processing. Parallel computing, uses multiple processing elements simultaneously to solve a problem. This is accomplished by breaking the problem into independent parts so that each processing element can execute its part of the algorithm simultaneously with the others. The processing elements can be diverse and include resources such as a single computer with multiple processors, several networked computers, specialized hardware, or any combinations of the above. Many mathematical computation applications involve multiple segments of code, some of which are repetitive. Often there is use of for-loops to solve these cases. The ability to execute code in parallel, on one computer or on a cluster of computers, can significantly improve performance for many such cases [6]. For example, a loop of 100 iterations could run on a cluster of 20 workers; so that simultaneously, each workers execute only five iterations of the loop. Here we may not get quite 20 times improvement in speed because of communications overhead and network traffic, However the speedup is significant. Parallelism are of different types, such as bit-level parallelism, instruction-level parallelism, data parallelism, task parallelism, etc. The work described here mainly focuses on data parallelism. Data parallelism is parallelism inherent in program loops, which focuses on distributing the data across different computing nodes to be processed in parallel (Fig. 10.1).

10.2.3 DSP Processor

Speech and image data requires a lot of convolutions, shift-add-accumulate operations for which DSP processors are suitable. DSP processors have certain advantages and have received attention in a vast range of areas of application covering communication, control, high-speed numeric processing, etc. The DSP processors have gained increased popularity because of the various advantages like reprogram ability in the field, cost-effectiveness, speed, energy efficiency, etc. [7].

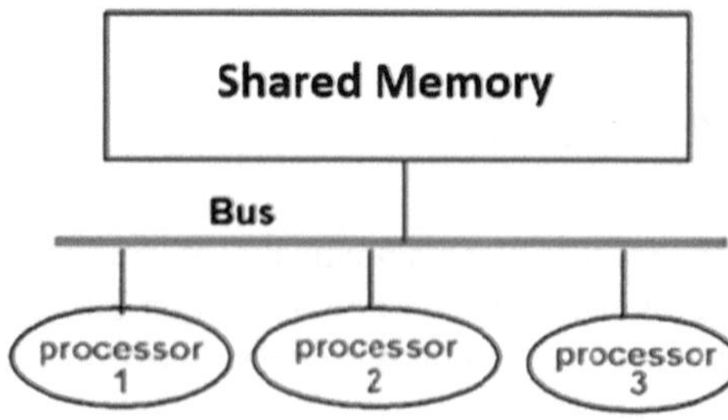

Fig. 10.1 Basic parallel processing architecture

The DSP processors are designed and optimized for implementation of various DSP algorithms. Most of these processors share various common features so as to support the high performance, repetitive, numeric intensive tasks, and lowering the computational complexity. Some of the key advantages are:

1. CPU Architecture: The C67x CPU is divided into two parallel parts, Data path A and Data path B. With this architecture, it is possible to process both left and right channels of an audio signal at the same time. The CPU is constructed using Harvard architecture where there are separate storage and signal pathways for instructions and data. Due to this, DSP able to fetch an instruction while simultaneously fetching operands and/or storing the result of previous instruction to memory. Some DSPs (for example TMS320C6XX) also include a small bank of RAM near the processor core, termed as L1 memory, which is used as an instruction cache. When a small group of instructions are executed repeatedly, the cache is loaded with these instructions thus making the bus available for data fetches, instead of instruction fetches.
2. MACs and Multiple Execution Units: This is the most commonly known segment of a DSP processor. It has the ability to perform one or more multiply and accumulate operation (also called as MAC) in a single instruction cycle. The MAC operation becomes useful as the DSP applications typically have very high computational requirements in comparison to other types of computing tasks, since they often execute DSP algorithms (such as FIR filtering) in real time on signals sampled at 10–100 kHz or higher.
3. Efficient Memory Access: DSP processors are also able to complete several accesses to memory in a single instruction cycle. The processor is able to fetch an instruction, while simultaneously fetching operands and/or storing the result of previous instruction to memory.
4. Circular Buffering: Circular buffers are used to store the most recent values of a continually updated signal. Circular buffering allows processors to access a block of data sequentially and then automatically wrap around to the beginning address exactly the pattern used to access coefficients in FIR filter.
5. Dedicated Address Generation Unit: The dedicated address generation units also help speedup the performance of the arithmetic processing on DSP. DSP processor addresses generation units typically and supports a selection of addressing modes tailored to DSP applications. The most common of these is register-indirect addressing with post-increment, which is used during the repetitive computation on data stored sequentially in memory. Some processors also support bit-reversed addressing, which increase the speed of fast Fourier transform (FFT) algorithms.
6. Specialized Instruction Sets: These sort of instruction sets of the digital signal processors maximize the utilization of the DSPs' resources ensuring the maximum efficiency and minimizing the storage space ensures the cost effectiveness of the overall system. Some of the latest processors now use VLIW (very long instruction word) architectures, where multiple instructions are issued and executed per cycle.

10.2.3.1 TMS320C6713 DSK

DSP processors such as the TMS320C6x (C6x) family of processors are fast special-purpose microprocessors with a specialized type of architecture and an instruction set appropriate for signal processing. The C6x notation is used to designate a member of Texas Instruments (TI) TMS320C6000 family of digital signal processors. The architecture of the C6x digital signal processor considered suitable for digital signal processing. Based on a VLIW architecture, the C6x is considered to be one of the TIs most powerful processor [8]. The TMS320C6713 DSK which has been used during the experimental work contains the TMS320C6713 digital signal processor. TMS320C6713 is a high performance floating-point DSP, its working frequency is up to 225 MHz, the single instruction execution cycle is only 5 ns, with a strong fixed-point; floating-point computing power generates a computational speed of up to 1.3 GFLOPS. TMS320C6713 processor consists of three main components: CPU core, memory, and peripherals. The CPU contains eight functional units, that can operate in parallel, two sets of registers, and address are 32 b wide. On-chip program memory bus has a width of 256 b. Peripherals include the expansion of the direct memory access (EDMA), low-power, external memory interface (EMIF), serial port, McBSP Interface, IIC interfaces, and timers. The C6713 DSK is a low-cost standalone development platform that enables users to evaluate and develop applications for the TI C67xx DSP family. Figure 10.2 shows the functional block diagram of the DSK. It also serves as a hardware reference design for the TMS320C6713 DSP. Schematics, logic equations, and application notes are available to ease hardware development and reduce time to market.

The DSK comes with a full compliment of on-board devices that suit a wide variety of application environments. Key features include [9]:

1. A Texas Instruments TMS320C6713 DSP operating at 225 MHz
2. An AIC23 stereo codec
3. 16 MB of synchronous DRAM
4. 512 kB of nonvolatile Flash memory (256 kB usable in default configuration)
5. Four user accessible LEDs and DIP switches

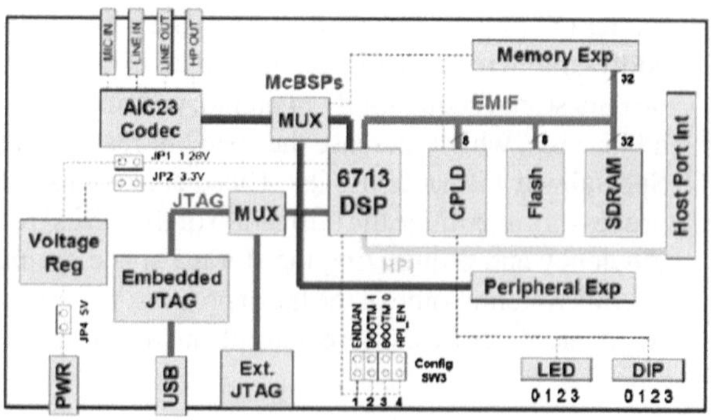

Fig. 10.2 Functional block diagram of TMS320C6713 DSK

6. Software board configuration through registers implemented in CPLD
7. Configurable boot options
8. Standard expansion connectors for daughter card use
9. JTAG emulation through on-board JTAG emulator with USB host
10. Interface or external emulator
11. Single voltage power supply (+5 V).

10.3 Result and Discussion

The ANN implementation to speech and image data is carried out using back propagation feedforward ANN algorithm. The samples, generated from different sources contain speech extracts and face captures. Some of the samples are mixed with noise. The sample sets thus generated consist of a sizeable number of data for use with the proposed system. Of these, about 25 % are categorized as training set, another 25 % for validation, and the rest taken for testing of the recognizer. Set of speech samples are recorded with variable sampling rates between 8 and 16 kbps. The soft computational framework is designed using the parallel computing environment by varying the numbers of CPU cores and DSP Processor. The performance comparison is then carried out. The work is carried out as per the process flow diagram, shown in Fig. 10.3. The ANN is made as per the configuration as shown in Table 10.1. For various number of hidden neurons, the number of epochs and processing time with TMS320c6713 and Intel Dual Core are shown in Tables 10.2 and 10.3 using patterns as speech data and image data, respectively.

Figure 10.4 shows the plot of MSE versus number of Epochs. It clearly dictates that TMS320C6713 process the ANN faster than the INTEL dual Core processor. As shown in Tables 10.2 and 10.3, for both the speech and image data, the number of epochs and hence the processing time required to meet the MSE of 1×10^{-3} is less for TMS320C6713, compared to the INTEL Dual Core and 4 core processor. For a number of hidden layer, neurons equal to (2× input layer) neurons the processing speed performance is found to be better compared to other cases of hidden layer

Fig. 10.3 Process flow diagram of the experimental setup

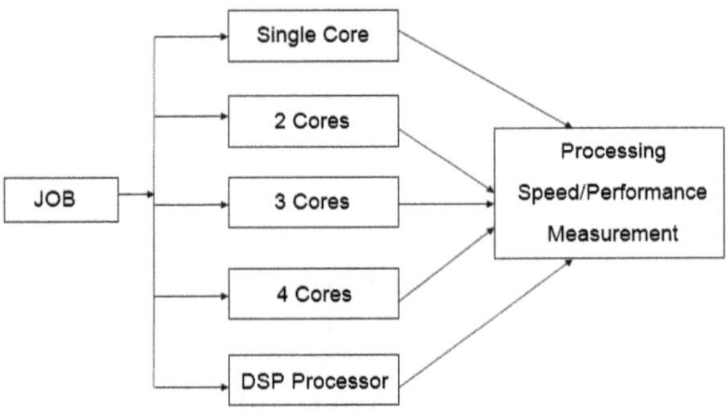

Table 10.1 Configuration of artificial neural network

Parameter	Value
No. of input neuron	50
No. of hidden neuron	Varied between 0.5 times and 2.5 times of the no of input neurons
No. of output neuron	4
No. of patterns	4
Transfer functions	Log sigmoid, tan sigmoid, log sigmoid
Learning rate	0.5

Table 10.2 Comparison of epochs and processing time for various number of hidden neurons with TMS320c6713 and Intel Dual Core using speech data pattern

No. of hidden layer neurons	Number of epochs required			Total time required (in s)		
	TMS320C6713	INTEL Dual Core	INTEL 4Core	TMS320C6713	INTEL Dual Core	INTEL 4Core
25	492	612	547	3.67	4.07	3.84
50	417	506	478	2.88	3.71	3.25
75	276	390	339	1.73	2.66	2.23
100	205	314	262	1.34	2.10	1.72
125	360	522	449	2.27	3.43	3.03

Table 10.3 Comparison of epochs and processing time for various number of hidden neurons with TMS320c6713 and Intel Dual Core using image data pattern

No. of hidden layer neurons	Number of epochs required			Total time required (in s)		
	TMS320C6713	INTEL Dual Core	INTEL 4Core	TMS320C6713	INTEL Dual Core	INTEL 4Core
25	549	719	667	3.45	3.91	3.63
50	361	594	510	2.19	3.28	2.82
75	185	403	326	1.30	2.82	2.57
100	148	336	261	1.05	2.41	2.14
125	395	458	402	2.74	3.67	3.06

neurons. In this case, compared to INTEL 4 Core processor, the TMS320C6713 processor provides improvement in processing speed efficiency of around 34 and 49 % for speech and image data, respectively. However, if we compare the processing time between the number of hidden layer neurons equal to (1.5× input layer) neurons and (2.5× input layer) neurons with reference to (2× input layer) neurons, we see

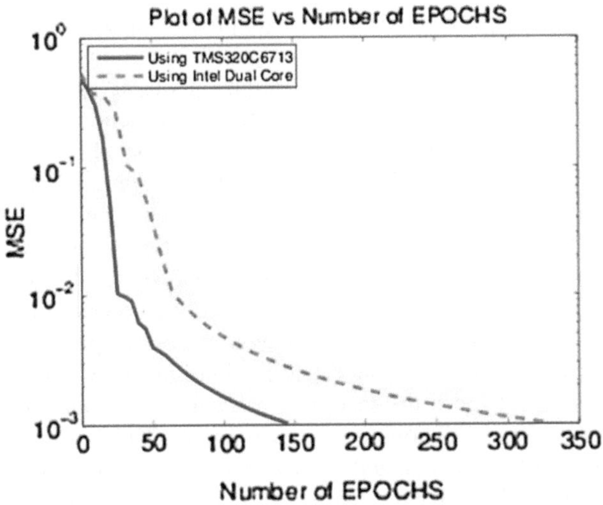

Fig. 10.4 Plot of MSE versus number of epochs for TMS320c6713 and Intel Dual Core

that the processing time for (2.5× input layer) neurons is much higher than (1.5× input layer) neurons. So this can be concluded here that for number of hidden layer neurons between (1.5× input layer) neurons to around (2× input layer) neurons, the soft-computational framework will have faster processing. Experimental results also shows that multicore CPU arrangement and the DSP processor-based framework helps ANN to learn applied patterns better.

10.4 Conclusion

Both speech and image signals require large storage capacity and processing for real-time applications and implementations. This dependence on storage and processing capability is more obvious in recognition systems based on learning. Among various classifiers, ANN is one of the prominent tools used for recognition of speech and image data. The ANN uses past knowledge for decision making using speech and image samples. DSP processor's high throughput characteristic and capability of executing million instructions per second provides better computational result as the ANN by design wise provides a parallel architecture. Similar characteristic is also shown by parallel processing environment. In this work, we investigated the performance of certain DSP processors and multicore parallel processing layouts for the design of a bio-inspired soft-computational framework. We demonstrated here the performance differences of ANN-based systems using speech and image inputs on multicore parallel processing layouts, DSP processor, and Intel duel core processor. We described the implementation of a parallel architecture with various numbers of processor cores for speech as well as image processing and compared their performances. The role played by parallel computing environment in increasing the processing performance of real-time applications involving speech processing and face recognition are also investigated in the work. Experimental results derived show that with multiple cores, parallel processing architectures aid the learning of ANN. It provides certain insights into bio-inspired system design and shows that multi-core CPU arrangement helps ANN to learn applied patterns better. We then used the DSP processors to design a bio-inspired soft-computational framework with which processing of speech and image inputs is carried out. The results derived show that the capability of the ANN improves with the derived DSP processor framework. Thus the formulated DSP-based framework for designing a bio-inspired soft-computation tool not only improves computational capability in terms of less number of processing cycles, but also enhances the recognition accuracy in both speech and image samples compared to parallel computing environment. This is obviously due to the support obtained from a specialized hardware like the DSP processor. It is thus obvious that the use of specialized hardware framework improves performance of soft-computational tools and facilitate design of bio-inspired system.

References

1. Seiffert U (2002) Artificial neural networks on massively parallel computer hardware. In: Proceedings of European symposium on artificial neural networks (ESANN), Bruges (Belgium), pp 319–330
2. Boser BE, Sackinger E, Bromley J, Cun YL, Jackel LD (1992) Hardware requirements for neural network pattern classifiers. A case study and implementation, IEEE Micro, pp 32–40
3. Asanovic K (2002) Programmable neurocomputing, The handbook of brain theory and neural networks, 2nd edn. The MIT Press. http://www.eecs.berkeley.edu/krste/papers/neurocomputing.pdf
4. Haykin S (2003) Neural networks a comprehensive foundation, 2nd edn. Pearson Education, New Delhi
5. Spetka SE, Ramseyer GO, Fitzgerald D, Linderman RE (2002) A distributed parallel processing system for command and control imagery. In: Proceedings of the command and control research and technology symposium
6. Grimaldi D, Rapuano S, Vito LD (2007) Parallel computing toolbox 5. Matlab User's Guide
7. Chassaing R (2005) Digital signal processing and applications with the C6713 and C6416 DSK. Wiley, Hoboken
8. Nadiminti K, Dias de AM, Buyya R (2006) Distributed systems and recent innovations: challenges and benefits. Grid Comput Distrib Syst Lab InfoNet Mag 16(3):1–5. http://www.gridbus.org
9. SPRS186L (2005) TMS320C6713 floating point digital signal processor. Texas Instruments

Part IV
Soft Computing and Hybrid System Based Speech Processing Applications

Three works constitute the section. These are by Talukdar et al., Goswami et al., and Misra et al. In the work by Talukdar et al. the authors present a comparative analysis of neuro-fuzzy approaches for clustering of speech data. The authors highlight the importance of clustering techniques as part of phoneme-based speech recognition. In the next chapter by Goswami et al., a speech signal reconstruction method using empirical mode decomposition (EMD) is described. It shows the use of EMD as an effective means of signal reconstruction as part of speech synthesis systems. The work of Misra et al. shows how vowel recognition of Assamese speech can be done using ANN and GMM approaches.

Chapter 11
Comparative Analysis of Neuro-Fuzzy Based Approaches for Speech Data Clustering

Pallabi Talukdar, Mousmita Sarma and Kandarpa Kumar Sarma

Abstract In this paper, we present a comparison between a few clustering algorithms including K-means clustering (KMC), Artificial Neural Network (ANN)-based Self-Organization Map (SOM), and Fuzzy C-means (FCM) clustering for the determination of number of phonemes present in a spoken Assamese word. Here, a block is designed to determine the number of phonemes present in a particular speech dataset. The phoneme count determination technique takes some initial decision about the possible number of phonemes present in a particular word. Comparing the success rate of correct decision from the proposed clustering techniques it is observed that the SOM-based technique provides more correct decisions compared to KMC-based technique and FCM-based approach provides even better decisions than the SOM-based technique. The results show that FCM generates better performance for all the cases considered.

Keywords Self-Organization Map (SOM) · Fuzzy C-means (FCM) · K-means Clustering (KMC) · Phonemes · Clustering

11.1 Introduction

Clustering is grouping of similar objects. The resulting groups are called clusters. Clustering algorithms group data points according to various criteria. Unlike most classification methods, clustering handles data that has no labels. The concept

P. Talukdar (✉) · M. Sarma
Department of Electronics and Communication Engineering,
Gauhati University, Guwahati, Assam, India
e-mail: pallabiz95@gmail.com

K.K. Sarma
Department of Electronics and Communication Technology,
Gauhati University, Guwahati 781014, Assam, India
e-mail: kandarpaks@gmail.com

M. Sarma
e-mail: go4mou@gmail.com

K.K. Sarma et al. (eds.), *Recent Trends in Intelligent and Emerging Systems*,
Signals and Communication Technology, DOI 10.1007/978-81-322-2407-5_11

mostly utilizes geometric principles, where the samples are interpreted as points in a d-dimensional Euclidian space, and clustering is made according to the distances between points [1]. Usually, points which are close to each other are assigned to the same cluster. The main concern in the clustering process is to reveal the organization of patterns into sensible groups, which allow us to discover similarities and differences, as well as to derive useful inferences about them. Clustering makes it possible to look at properties of whole clusters instead of individual objects, i.e., a simplification that might be useful when handling large amount of data. It has the potential to identify unknown classification schemes that highlight relations and differences between objects [2]. The most fundamental clustering technique is the KMC. Its objective is to minimize the average squared Euclidean distance of documents from their cluster centers which are statistical means or centroids. Clustering has been used for a range of applications including classifier design. Artificial Neural Networks (ANNs) are useful techniques for clustering where there are two primary stages. In the first, the learning rule is used to train the network for a specific dataset. In the second stage, the observations are classified, which is called a recall or testing stage [3].

The K-means algorithm is a simple iterative clustering algorithm that partitions a given dataset into a user-specified number of clusters, K. The algorithm is simple to implement and run, relatively fast, easy to adapt, and common in practice. Self-Organization Map (SOM) is one of the most popular ANN models for unsupervised learning [4–6]. The name SOM signifies a class of ANN algorithms in the competitive learning category. SOM has been successfully applied in clustering and visualization of high-dimensional data [7, 8]. However, application of the standard SOM algorithm relies on the validity of the squared Euclidean distance between the map units and data as an error function [9, 10]. Another cluster technique called the Fuzzy C-means (FCM) uses the advantages of fuzzy system. FCM clustering is an unsupervised technique that has been successfully applied to feature analysis, clustering, and classifier designs in fields such as astronomy, geology, medical imaging, and pattern recognition [11, 12].

In this work, the phoneme cluster determination technique with special stress to Assamese language is designed using the above-mentioned three different clustering approaches. The clustering methods namely KMC, SOM, and FCM are used along with a Recurrent Neural Network (RNN) block for decision-making. The KMC-based technique provides around 85 % correct decision with respect to the number of phonemes. The SOM clustering-based phoneme count determination technique provides superior performance in terms of percentage of correct decision than the KMC. However, the results shown by the FCM is the best among the three considered and is consistent for all the cases considered. Assamese consonant-vowel-consonant (CVC) words are recorded for this work in a noise-free environment from five male and five female speakers of varying age speaking a number of CVC combinations. These are used for training, validation, and testing of the proposed method. Though clustering has been used for a range of applications, no work has been reported to the best of our knowledge, which considers KCM, SOM, and FCM-based clustering for phoneme extractions in spoken words and that too in a Devanagari-based language like Assamese.

The description included here is organized as below. Section 11.2 provides briefly the proposed model for determination of phonemes which consists of following parts—KMC-based clustering algorithm, SOM-based clustering algorithm, and FCM-based clustering algorithm. The results and the related discussions are included in Sect. 11.3. Section 11.4 concludes the description.

11.2 Proposed Model for Determination of Phoneme Cluster

The proposed phoneme cluster determination model can be described by the block diagram of Fig. 11.1. This model can be categorized into three parts as follows:

- KMC-based clustering algorithm,
- SOM-based clustering algorithm, and
- FCM-based clustering algorithm

The phoneme cluster determination block determines the number of phonemes present in a particular word. Comparing the success rate of correct decision from the

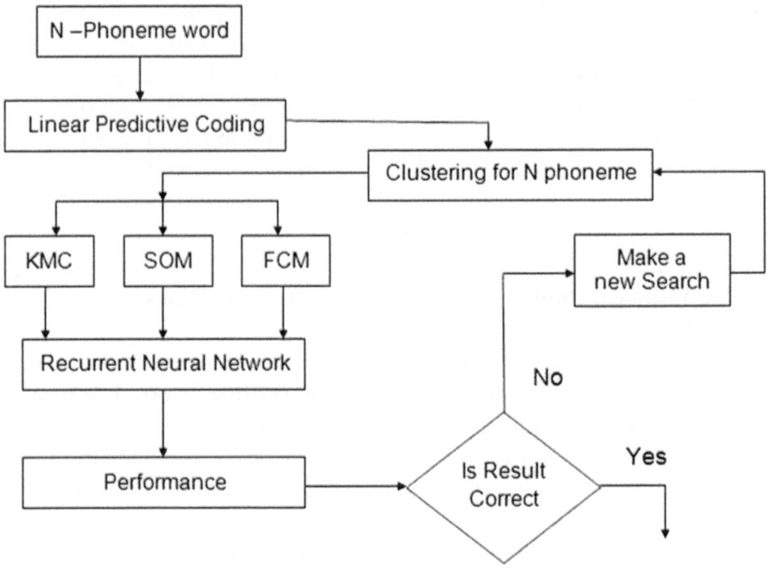

Fig. 11.1 Phoneme cluster determination block

above-mentioned algorithms, it is observed that the SOM-based algorithm can provide more correct decisions compared to KMC-based algorithm and FCM-based algorithm provides more correct decisions than the SOM-based algorithm. The results are shown in Sect. 11.3.

11.2.1 Phoneme Cluster Determination

The phoneme cluster determination technique takes some initial decision about the possible number of phonemes present in a particular word. The phoneme cluster determination block determines the number of phonemes present in a particular word. From the incoming words having different phonemes, initially the LPC feature is extracted and prepared for clustering and presented to the RNN for classification [13]. The RNN is trained to learn number of clusters in the data.

11.2.1.1 KMC-Based Phoneme Cluster Determination Block

K-means clustering is an elementary but popular approximate method that can be used to simplify and accelerate convergence [14]. Clustering is the process of partitioning or grouping a given set of patterns into disjoint clusters. This is done such that patterns in the same cluster are alike and patterns belonging to two different clusters are different [15, 16]. Clustering has been a widely studied problem in a variety of application domains including ANNs. The K-means algorithm is a simple iterative clustering algorithm that partitions a given dataset into a user-specified number of clusters, K. The number of clusters K is assumed to be fixed in K-means clustering. The method involves clustering N-phoneme words into N-clusters using KMC. Here, N is the value of K in the KMC. The word clustering technique proposed here can be visualized in Fig. 11.2. The KMC algorithm steps can be stated as follows (Table 11.1).

11.2.1.2 SOM-Based Phoneme Cluster Determination Block

The same phoneme cluster determination block is later redesigned using the SOM. Here, the KMC blocks are replaced by SOM competitive layers. The basic idea of a SOM is to map the data patterns onto a n-dimensional grid of neurons or units. That grid forms the output space, as opposed to the input space where the data patterns are applied. This mapping preserves topological relations, i.e., patterns that are close in the input space will be mapped to units that are close in the output space, and vice versa [15]. The SOM training algorithm resembles K-means algorithm in the sense that it partitions the input data space into a number of clusters of wining neuron and its neighbors with similar weight vectors. Therefore, SOM can be used

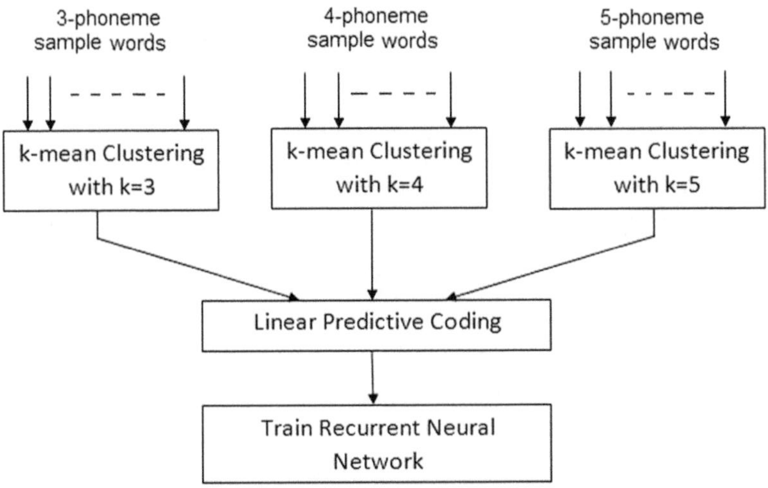

Fig. 11.2 Word clustering technique using KMC

Table 11.1 KMC algorithm

1. Input: Speech S of size $m \times n$, sampling frequency f_s, duration T second;
2. Initialize N=100, order of linear prediction;
3. Apply K-means clustering on input dataset S such as $s^{(1)}, s^{(2)} \ldots . s^{(m)}$;
4. Set the number of clusters K ;
5. Initialize cluster centroids as $\mu_1, \mu_2 \ldots . \mu_k$.
6. Repeat until convergence
   For every i, set

$$C^{(i)} = argmin \, || \, s^{(i)} - \mu_j \, ||^2, \tag{11.1}$$

7. Update centroid position;
8. Compute new centroid position from assigned members.

for clustering data without knowing the class memberships of the input. The SOM algorithm is based on unsupervised, competitive learning, where clusters are formed depending upon a self-organization process of the constituent neurons which groups data depending upon a similarity measure. The similarity measure is decided upon by a Euclidian distance between the random connectionist values and the input and finally optimized by a Gaussian spread [17]. The SOM-based phoneme count determination technique can be summarized from the Fig. 11.3. The SOM algorithm steps can be stated as Table 11.2.

Table 11.2 SOM algorithm

1. Input: Speech S of size $m \times n$, sampling frequency f_s, duration T second;
2. Initialize N=100, order of linear prediction;
3. Take a SOM topology map of hexagonal pattern;
4. Apply SOM clustering on input dataset S such as $s^{(1)}, s^{(2)} \ldots s^{(m)}$;
5. Find the winning neuron $I(S)$ that has weight vector closest to the input vector, i.e.

$$\min d_j(x) = \sum_{i=1}^{D} (s_i - w_{ji})^2, \tag{11.2}$$

5. Apply weight update equation

$$\Delta w_{ji} = \eta(t) T_{j,I(S)}(t)(s_i - w_{ji}) \tag{11.3}$$

where
 $T_{j,I(S)}$(t) is a Gaussian neighbourhood,
 η(t) is learning rate and
 w_{ji} is weight vector.

6. Adjust the weights of the winning neuron and neighbouring neuron;
7. Keep returning to step 4 until the feature map stop changing.

11.2.1.3 FCM-Based Phoneme Cluster Determination Block

FCM adopts the advantages of fuzzy systems to provide better clustering. FCM clustering is dependent of the measure of distance between samples. In most situations, FCM uses the common Euclidean distance which supposes that each feature has equal importance in FCM [14]. FCM is a data clustering technique wherein each data point belongs to a cluster to some degree that is specified by a membership grade. It provides a method that shows how to group data points that populate some multi-dimensional space into a specific number of different clusters [17]. The FCM-based phoneme cluster determination technique can be summarized from the Fig. 11.4. The FCM algorithm steps can be stated as follows Table 11.3:

Table 11.3 FCM algorithm

1. Input: Speech S of size $m \times n$, sampling frequency f_s, duration T second;
2. Initialize N=100, order of linear prediction;
3. Apply FCM clustering on input dataset S such as $s^{(1)}, s^{(2)} \ldots s^{(m)}$;
4. Initialize number of cluster C;
5. Calculate centroid for speech data;
6. For each point, find its closest center;
7. Return the best centers found;
8. Repeat until convergence.

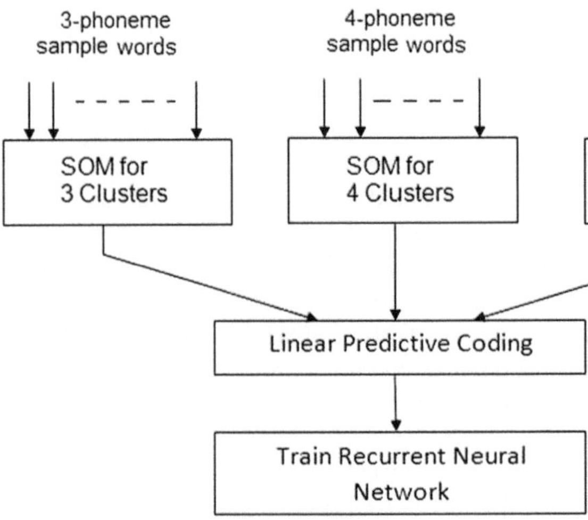

Fig. 11.3 Word clustering technique using SOM

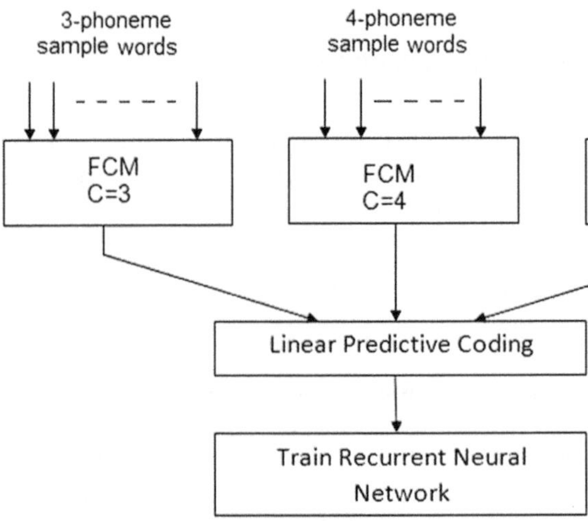

Fig. 11.4 Word clustering technique using FCM

11.2.1.4 Role of RNN in Decision-Making of the Proposed Technique

RNN is a special structure of supervised ANN known for temporal data processing and is suited for capturing the time-varying nature of speech signals. Therefore, it has been chosen for taking decision about the number of clusters in an Assamese CVC speech data. An RNN is trained with a composite speech signal set enable the learning from which cluster count is extracted and subsequent classification decision provided. The sample clustered dataset is obtained from KMC, SOM or FCM, where

3-phoneme words have 3 clusters, 4-phoneme words have 4 clusters, and 5-phoneme words have 5 clusters. Accordingly, the RNN classifies the data into 3, 4, and 5-cluster groups. Suppose, they are stored as DK3, DK4, and DK5 respectively. Here, we have considered only certain variations like 3-, 4-, and 5- phonemes. Now, any N-phoneme word obtained from the discrete speech signal is first clustered with $K = 3, 4, 5$. Next, at first DK3 is presented to the trained RNN to classify into any of the three classes. If RNN fails to classify, then DK4 and DK5 are presented consequently. If any of DK3, DK4, and DK5 do not come under any of the classes defined by RNN, then that word is discarded. The same decision-making method is used in case of clustered dataset obtained from KMC-, SOM-, and FCM-based clustering techniques. Comparing the success rate of correct decision from the above-mentioned techniques, it is seen that the SOM-based technique provides more correct decisions (around 7 % improvement) than the KMC-based technique. However, FCM-based technique shows more correct decisions (around 3 % improvement) than the SOM-based technique. The results of the three different techniques are described further in Sect. 11.3.1

11.3 Experimental Result and Discussion

The experimental speech samples are CVC-type Assamese words recorded from five female speakers and five male speakers of varying ages. The recorded speech samples has the following specification:

- Duration = 2 s,
- Sampling rate = 8000 samples/s,
- Bit resolution = 16 bits/sample.

Selection of proper set of sample words as a representative of all the N-phoneme words, where (N $=$ 2, 3 or 4) is an important factor for better success rate of the proposed technique. The RNN should be trained with enough data, so that it can learn the difference between 3-phoneme, 4-phoneme, and 5-phoneme words. The work described in this paper, experimented the proposed technique only for 3-phoneme, 4-phoneme, and 5-phoneme words. Accordingly, a word list is prepared for clustering purpose. Such a sample word list is shown in Table 11.4. Around 150 words are recorded for each of the male and female speakers. Of this, 30 % of the words are used for training, 40 % are used for validating, and 30 % are used for testing the clustering technique. The dataset is further increased by adding Gaussian noise between -3 and $+3$ dB. Also, two samples of speech word having 3-phoneme, 4-phoneme, and 5-phoneme from each class of prepared speech dataset are shown in Fig. 11.5. The following sections provide the result obtained by the proposed phoneme cluster determination block.

Table 11.4 Word list prepared for clustering technique

Sl. no.	Word	Correct decision	False decision
1	/kam/	/akal/	/kakai/
2	/kar/	/ekal/	/kakal/
3	/dam/	/asal/	/kasam/
4	/kath/	/aghat/	/kadam/
5	/kat/	/usar/	/kapal/
6	/xar/	/asin/	/kapah/
7	/sar/	/azay/	/kamal/
8	/dhar/	/azat/	/kamar/
9	/nar/	/atal/	/katar/
10	/kan/	/katha/	/kasam/
11	/mok/	/dora/	/betal/
12	/mor/	/beli/	/bixal/
13	/mot/	/mora/	/muk/
14	/mon/	/mona/	/xapon/
15	/bel/	/mula/	/xiphal/
16	/xir/	/besi/	/natun/
17	/xik/	/bora/	/labar/
18	/xis/	/dili/	/lagar/
19	/bes/	/lora/	/nagar/
20	/bed/	/nila/	/nazar/

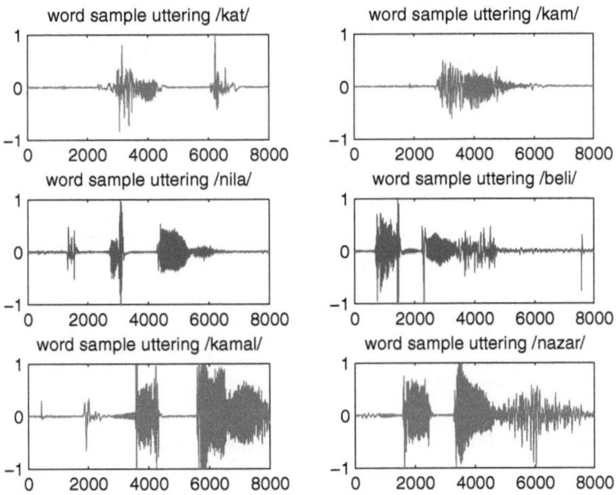

Fig. 11.5 Speech samples of 3-phoneme, 4-phoneme, and 5-phoneme

11.3.1 Comparative Performance of KMC, SOM, and FCM Clustering Methods

From the incoming words having different phonemes, initially the LPC feature is extracted and clustered with KMC algorithm for K-value 3. The clustered data is then presented to the RNN for classification. The RNN is trained to learn number of clusters in the data. If RNN fails to classify the data into any of the defined class, then the word is clustered with KMC algorithm for K-value 4 and the process is repeated. If the RNN again fails to classify the data then the word is clustered with K-value 5 and the same process is repeated again. In this way, the proposed logic is used to determine the possible number of phonemes in the word for 3-, 4- and 5-phoneme words. If the RNN fails to take any decision, then that particular word is discarded. Selection of proper set of sample words as a representative of all the N-phoneme words, where (N = 2, 3 or 4) is an important factor for better success rate of the proposed technique. The RNN should be trained with enough data, so that it can learn the difference between 3-, 4-, and 5-phoneme words. Here, we have considered words having all the Assamese vowel variations. Words with all the possible vowel and consonant combination of Assamese vocabulary, if are used for the purpose, shall provide better results. The samples classified, as mentioned above, are used for testing the KMC technique. From the experimental results, it is seen that the KMC-based method can give around 85 % success rate, while determining number of phonemes in a word using KMC-aided apriori knowledge for noise-free signals and it is shown in Table 11.5. Next, the cluster block using KMC is replaced by SOM. The SOM-based phoneme cluster determination block is designed as explained in Sect. 11.2.1.2. The success rate phoneme cluster determination using SOM is shown in Tables 11.6 and 11.7 for noise-free and noisy signal respectively. As described already, a noisy signal set is created by adding −3 dB to +3 dB Gaussian noise with the noise free signals. It can be seen that the noise-free signal set shows around 7 %

Table 11.5 Phoneme cluster determination success rate for noise-free signals using KMC

Sl. no.	Word	Correct decision (%)	False decision (%)
1	3-Phoneme	85	6
2	4-Phoneme	88	8
3	5-Phoneme	81	8

Table 11.6 Phoneme cluster determination success rate for noise-free signals using SOM

Sl. no.	Word	Correct decision (%)	False decision (%)
1	3-Phoneme	92	6
2	4-Phoneme	89	8
3	5-Phoneme	90	8

Table 11.7 Phoneme cluster determination success rate for noisy signals using SOM

Sl. no.	Word	Correct decision (%)	False decision (%)
1	3-Phoneme	84	10
2	4-Phoneme	86.2	8.5
3	5-Phoneme	87	6.2

Table 11.8 Phoneme cluster determination success rate for noise-free signals using FCM

Sl. no.	Word	Correct decision (%)	False decision (%)
1	3-Phoneme	95	5
2	4-Phoneme	90	10
3	5-Phoneme	93	7

Table 11.9 Phoneme cluster determination success rate for noisy signals using FCM

Sl. no.	Word	Correct decision (%)	False decision (%)
1	3-Phoneme	88	8.5
2	4-Phoneme	82.6	10
3	5-Phoneme	80	8.3

Table 11.10 Success rate in % with three different clustering approaches

	Phoneme type	KCM (%)	SOM (%)	FCM (%)
Noise free	3-Phoneme	85	92	95
	4-Phoneme	88	89	90
	5-Phoneme	81	90	93
Noise mixed	3-Phoneme	67.5	84	88
	4-Phoneme	72	86.2	82.6
	5-Phoneme	70	87	80

improvement in comparison to the KMC-based method. Further increasing the RNN training signals the success rate of the noisy signal set can also be improved. The success rate of the phoneme cluster determination using FCM is shown in Tables 11.8 and 11.9 for noise-free and noisy signal.

Comparitive success rates of KMC, SOM, and FCM clustering methods for noise-free and noisy samples are shown in Table 11.10. From the Table 11.10 it is seen that the noise-free signal set shows around 7 % improvement in case of SOM-based method. Again it is also seen that around 3 % improvement can be seen in comparision to the SOM-based method. The results showed that FCM provides better performance in all cases because FCM has capability to tract very minute variation present in the

Table 11.11 Average computational time of RNN using KMC

Cluster method	Phoneme type	Training time	Testing time (s)
	3-Phoneme		2.15–2.68
KMC	4-Phoneme	27.82 s	2.32–2.59
	5-Phoneme		2.51–2.91

Table 11.12 Average computational time of RNN using SOM

Cluster method	Phoneme type	Training time	Testing time (s)
	3-Phoneme		2.85–3.12
SOM	4-Phoneme	31.33 s	2.91–3.25
	5-Phoneme		3.11–3.45

Table 11.13 Average computational time of RNN using FCM

Cluster method	Phoneme type	Training time	Testing time (s)
	3-Phoneme		1.98–2.15
FCM	4-Phoneme	25.67 s	1.68–1.85
	5-Phoneme		1.72–1.90

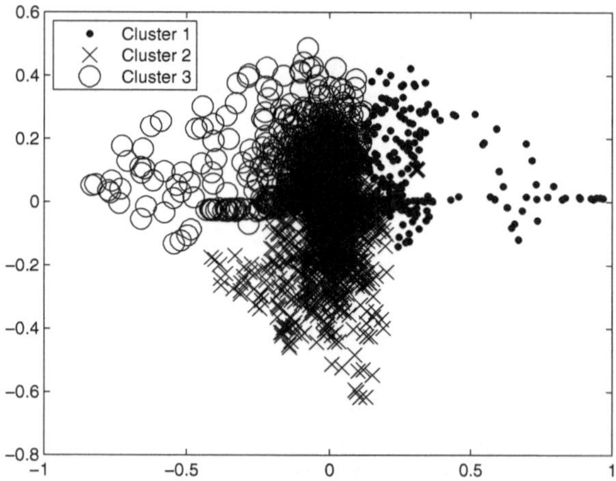

Fig. 11.6 Clustering using KMC for 3-phoneme

speech data. Also, computational time of RNN by using the three clustering methods are shown in Tables 11.11, 11.12 and 11.13 respectively. The SOM-based approach takes maximum time during training but its success rate is a bit lower than that shown by FCM. The FCM takes around 25 s to train while providing cluster outputs. It also has higher success rates compared to SOM and KMC. Thus, the FCM-based method

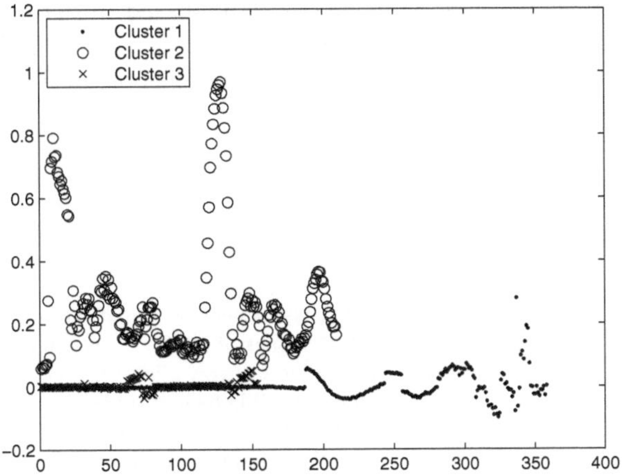

Fig. 11.7 Clustering using SOM for 3-phoneme

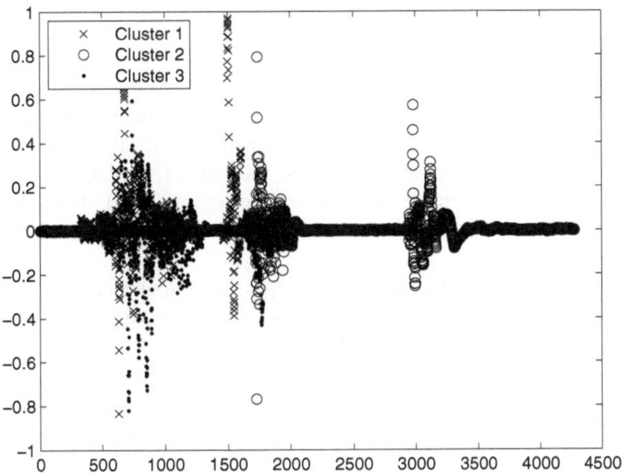

Fig. 11.8 Clustering using FCM for 3-phoneme

is a suitable one for the providing cluster information as a part of a RNN-aided recognition mechanism.

Here, the clustering of speech data for 3-phoneme using above three methods are shown in Figs. 11.6, 11.7 and 11.8 respectively.

11.4 Conclusion

In this work, we have described a phoneme cluster determination block. The phoneme cluster determination block counts the number of phonemes clusters present in a particular word. Comparing the success rate of correct decision from the proposed clustering method, it is observed that the SOM-based clustering method can provide more correct decisions compared to KMC-based clustering method and FCM-based clustering method provides even better correct decisions than the SOM-based clustering method. Thus, a FCM-based clustering approach is suitable for phoneme clustering application as part of a speech recognition system.

References

1. Fukuyama Y, Sugeno M (1989) A new method of choosing the number of clusters for the fuzzy C-means method. In: Proceedings of 5th fuzzy system symposium, pp. 247–250
2. Granath G (1984) Application of fuzzy clustering and fuzzy classification to evaluate provenance of glacial till. J Int Assoc Math Geol 16(3):283–301
3. Mingoti AS, Lima OJ (2006) Comparing SOM neural network with fuzzy C-means, K-means and traditional hierarchical clustering algorithms. Eur J Oper Res 174(3):17421759
4. Kohonen T (1990) The self-organizing map. Proc IEEE 78(9):547–560
5. Bar GD, Kuflik T, Lev D (2009) Supervised learning for automatic classification of documents using self-organizing maps. Proc IEEE Int Conf 2:1136–1139
6. Kohonen T (1993) Generalization of the self-organizing map. In: Proceedings of international joint conference on neural networks, pp. 457–462
7. Ultsch A, Vetter C (2001) Self-organizing feature maps versus statistical clustering methods: a benchmark. http://citeseerx.ist.psu.edu/viewdoc/download?doi=10.1.1.196.7322
8. Kohonen T, Kaski S, Lagus K, Salojarvi J, Honkela J, Paatero V, Saarela A (2000) Self organization of a massive document collection. IEEE Trans Neural Networks 11(3):574–585
9. Allinson N, Yin H, Allinson L, Slack J (2001) Advances in self-organising maps. Springer, London
10. Sarma M, Sarma KK (2012) Segmentation of assamese phonemes using SOM. In: Proceedings of 3rd IEEE national conference on emerging trends and applications in computer science (NCETACS). Shillong, India, pp 121–125
11. Iyer SN, Kandel A, Schneider M (2002) Feature-based fuzzy classification for interpretation of mammograms. Fuzzy Sets Syst. doi:10.1016/S0165-0114(98)00175-4
12. Wang X, Wang Y, Wang L (2004) Improving fuzzy C-means clustering based on feature-weight learning. Pattern Recogn Lett 25(10):11231132
13. Sarma M, Sarma KK (2012) Recognition of assamese phonemes using three different ANN structures. In: Proceedings of CUBE international IT conference. Pune, India, pp 121–125
14. Duda OR, Hart EP, Stork GD (2000) Pattern classification, 2nd edn. Wiley-Interscience Publication, New York
15. Haykin S (2009) Neural network and learning machine, 3rd edn. PHI Learning Private Limited, New Delhi
16. Teknomo K (2003) K-means clustering tutorial. IEEE Press
17. Hu HY, Hwang NJ (2002) handbook of neural network signal processing. The electrical engineering and applied signal processing. CRC Press, Boca Raton

Chapter 12
Effective Speech Signal Reconstruction Technique Using Empirical Mode Decomposition Under Various Conditions

Nisha Goswami, Mousmita Sarma and Kandarpa Kumar Sarma

Abstract Empirical mode decomposition (EMD) is a method for nonstationary signal analysis, where signals are decomposed into number of high-frequency modes called intrinsic mode function (IMF)s and a low-frequency component called the residual. If this residual is considered as the source signal in case of a speech signal and vocal tract filter response is estimated, the original signal can be reconstructed. The nonstationary attribute of a speech signal restricts direct application of the conventional digital signal processing (DSP) techniques to a speech signal. Since EMD performs decomposition assuming the nonstationary nature of a signal, it has been observed that frame-by-frame analysis is not required in the proposed reconstruction model. The effectiveness of reconstruction using EMD residual is experimented in case of speech data collected in various conditions like clean, noisy, mobile channel including speaker's mood variation.

Keywords Empirical mode decomposition (EMD) · Intrinsic mode function (IMF)s · Speech signal · Reconstruction

12.1 Introduction

Glottal source and vocal tract filter separation is a crucial component of most of the applications in speech processing. Like almost all natural phenomena, speech is the result of many nonlinearly interacting processes.Therefore, any linear analysis has the potential risk of underestimating, or even missing, a significant amount

---

N. Goswami (✉) · K.K. Sarma
Department of Electronics and Communication Technology, Gauhati University, Guwahati 781014, Assam, India
e-mail: nisha.goswami75@gmail.com

K.K. Sarma
e-mail: kandarpaks@gmail.com

M. Sarma
Department of Electronics and Communication Engineering, Gauhati University, Guwahati, Assam, India
e-mail: go4mou@gmail.com

K.K. Sarma et al. (eds.), *Recent Trends in Intelligent and Emerging Systems*, Signals and Communication Technology, DOI 10.1007/978-81-322-2407-5_12

of information content [1]. Time-frequency representation is essential in signal processing and has the potential of diverse applications. Examples include short-time Fourier transform (STFT), wavelet transform, etc. Most of the digital signal processing techniques assume that the data is linear and stationary [2]. Speech is usually both nonlinear and nonstationary. The empirical mode decomposition (EMD), first introduced by N.E. Huang et al. in 1998, is designed to reduce nonstationary, multicomponent signals to a series of amplitude and frequency modulation (AM and FM) contributions. The net result is to create a bank of subsignals, termed intrinsic mode function (IMF)s whose sum produces the original signal. The last IMF or residual is of the lowest order [3]. The residual information provides us the information of source. Due to the nonstationary assumption of EMD, it is suitable for temporal data analysis like speech.

Some works have been reported which explores EMD for speech processing applications. In [4], the reported work illustrates a novel and effective method for suppressing residual noise from enhanced speech signals as a second-stage post-filtering technique using EMD. In [5], the authors show that the speech fundamental frequency can be captured in a single IMF. Thus a new algorithm for pitch extraction based on the ensemble empirical mode decomposition (EEMD) is presented [5]. In [6], EMD is used for extraction of long-term structures in musical signals, which provide information concerning rhythm, melody, and the composition. In this chapter, an EMD-based approach is described for speech signal reconstruction application.

EMD is applied for extracting the glottal source information of speech signals. After getting the source information, vocal tract filter response is determined and the original speech signal is reconstructed. The experimental result derived establishes the effectiveness of the proposed method.

The rest of the paper is organized as below:

Sect. 12.2 describes about the speech signal reconstruction using EMD. The experimental details and results are described in Sects. 12.3 and 12.4 includes the conclusion and future direction.

12.2 Speech Signal Reconstruction Using EMD

In this section, we have discussed about EMD technique and its usefulness to extract residual content from speech signal and carry out reconstruction of original speech signal. EMD is an adaptive tool to analyze nonlinear or nonstationary signals which segregates the constituent parts of the signal based on the local behavior of the signal. Using EMD, signals can be decomposed into number of frequency modes called intrinsic mode function (IMF). An IMF represents a simple frequency mode similar to the simple harmonic function, but mode frequency has amplitude and frequency as function of time unlike simple harmonic component [2]. An IMFs satisfy two conditions. The first is that the number of extrema (sum of maxima and minima) and the number of zero-crossing must be equal or differ by one. The second is the mean of the cubic splines must be equal to zero at all points [7].

For speech signal $x(t)$, first step is identification of all the extrema (maxima and minima) of the series $x(t)$ and then generation of the upper and lower envelope via cubic spline interpolation among all the maxima and minima, respectively. Point-by-point averaging of the two envelopes is performed to compute a local mean series $m(t)$. In order to obtain a IMF, subtraction of $m(t)$ from the data is done, i.e., $h(t) = x(t) - m(t)$ and then the properties of $h(t)$ are checked to verify whether it is a IMF or not. If h is not a IMF, $x(t)$ is replaced by $h(t)$ and repeated. Again if h is an IMF, the residue is evaluated as $m(t) = x(t) - h(t)$. The above steps are repeated by shifting the residual signal, until at least two extrema remains. After getting the residual, fft of original signal $x(t)$ and $r_n(t)$ is taken. Later vocal tract transfer function in frequency domain $H(\omega)$ dividing $X(\omega)$ by Fourier transform of glottal source $R(\omega)$. Since now residual and vocal tract filter response is available; hence speech signal can be reconstructed easily from the concept of source filter model. Suppose, $r_n(t)$ is the source information and $x(t)$ be the known speech signal. Let $h(t)$ be the filter response in time domain. These can be related as

$$x(t) = r_n(t) * h(t) \tag{12.1}$$

Representing Eq. 12.1 in frequency domain, we get

$$X(\omega) = R(\omega)H(\omega) \tag{12.2}$$

Therefore,

$$H(\omega) = X(\omega)/R(\omega) \tag{12.3}$$

Equation 12.3 represents the vocal tract filter response. The pseudocode for the proposed model is given by the following steps given in Table 12.1 and the complete system model is shown Fig. 12.1. Speech production model can be represented as a source filter model as shown in Fig. 12.2. The speech sound wave and the vocal tract are modeled as a signal and a filter, respectively [8]. A speech signal is first presented to the EMD block to get the residual. It represents the glottal source information. Now, if frequency domain division of the speech signal by the glottal source is performed, then the vocal tract filter response can be obtained. The last step is the filtering of the residual signal by the estimated vocal tract filter. Vocal tract filter represents the signal processing equivalent of the biological system which gives the original speech signal. Thus, the EMD process provides the glottal source information and combines it together with the vocal tract filter.

In the present technique of speech signal reconstruction, initially EMD is applied to a known speech signal. EMD method is applicable for both nonlinear and nonstationary signals. Speech is not a stationary signal, i.e., it has properties that change with time. In general cases, data is assumed to be linear and digital processing techniques applied are like short-time Fourier transform (STFT). STFT is essentially composed of piecewise FFTs, which assumes that the signal of interest is locally stationary. But when EMD is used, it considers speech as nonstationary and applied

Table 12.1 EMD-based speech reconstruction algorithm

1. Initialize $: r_0(t) \longleftarrow x(t), i \longleftarrow 1$, where $x(t)$ be speech signal taken.
2. Extract the ith IMF.
   a. Initialise $h_0(t) \longleftarrow r_i(t)$,$i \longleftarrow 1$.
   b. Extract the local minima and maxima of $h_{j-1}(t)$.
   c. Interpolate the local maxima and local minima by a cubic spline to form upper and lower envelopes of $h_{j-1}(t)$.
   d. Calculate the mean $m_{j-1}(t)$ of upper and lower envelop.
   e. $h_j(t) \longleftarrow h_{j-1}(t)$-$m_{j-1}(t)$.
   f. If stopping criterion is satisfied then $IMF_i(t) \longleftarrow h_j(t)$
   else
   $j \longleftarrow j+1$ and goto (b).
3. $r_i(t) \longleftarrow r_{i-1}(t)$ - $IMF_i(t)$.
4. If $r_i(t)$ still has at least two extrema then $j \longleftarrow j+1$ and goto (b) else decomposition is finished and $r_i(t)$ is the residue.
5. After obtaining the residue, take fft of $x(t)$
6. Take fft of $r_n(t)$ (residual signal), where n is the number of IMFs.
7. Divide $X(\omega)$ by fourier transform of glottal source $R(\omega)$ to obtain vocal tract transfer function in frequency domain $H(\omega)$.
8. Reconstruction of original speech signal $x(t)$ with the help of vocal tract $h(t)$ and glottal source $r_n(t)$.

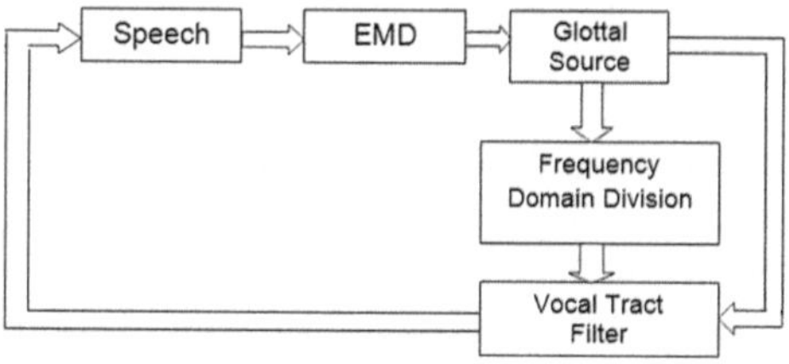

Fig. 12.1 System model

Fig. 12.2 Source filter model of speech production

on the whole signal rather using short-term processing methods. EMD has satisfactory computational efficiency and does not involve the concept of the frequency resolution and the time resolution. No preprocessing is required since it is able to analyze nonzero mean signals, and is suitable to analyze the riding waves which may have no zero-crossings between two consecutive extrema [6]. Since speech signals are nonlinear, EMD is applied to the speech signal to decompose it into a number of IMFs. It finally provides us the lowest frequency component, i.e., the residual signal. The residual signal provides us the source information. After finding the source information vocal tract filter response is determined as explained above.

12.3 Experimental Results and Discussion

Here, we briefly discuss the results derived for a range of experiments performed. We have recorded speech signals in microphone channel and mobile phone channel. The microphone channel data are collected with three different moods of the speaker like angry, loud, and soft. Further, we have added noise to the speech signals and applied the proposed EMD-based algorithm of speech reconstruction to these speech samples. The description of the experiments are as follows:

1. **Proposed model under normal conditions**:
   Initially, speech samples recorded under normal conditions are used for the proposed reconstruction model. Twenty different speech samples from 10 male and 10 female speakers are taken which are recorded in noise free office room using a headphone and speech analysis software wavesurfer. The recording specifications are as follows:

   - Bit rate—16 bit/sample
   - Sampling Frequency—48,000 samples/s

   The EMD-based speech reconstruction algorithm described in Sect. 12.2 is applied to the speech signals separately. The speech signal along with their 15 IMFs and residual for a male speaker are shown in Fig. 12.3 through Fig. 12.6, where Figs. 12.3, 12.4 and 12.5 represents IMF1 through IMF12 and Fig. 12.6 represents the IMF13, IMF14, and the residual signal. After getting the residual signal, vocal tract filter response is determined as discussed earlier and speech signal is reconstructed with the help of source and vocal tract filter response, which is found to be similar to the original speech signal after converting it to time domain using inverse Fourier transform. Figure 12.7 shows the original and reconstructed speech signal. The experiment is repeated for 10 male and 10 female

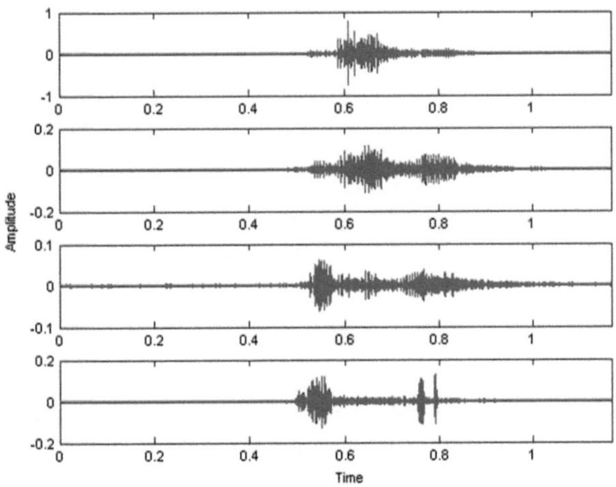

Fig. 12.3 IMF1 to IMF4 of a male speaker

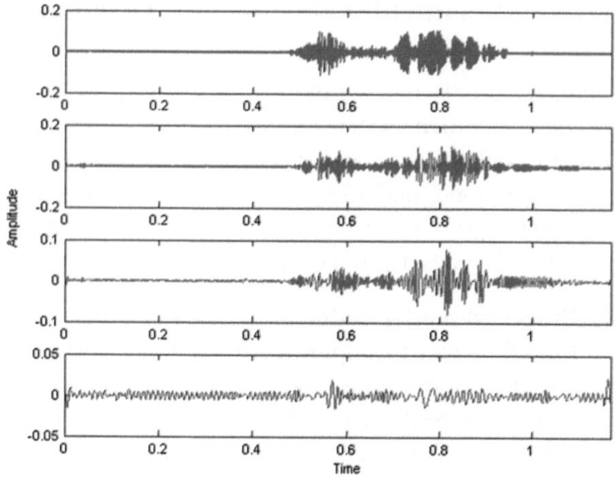

Fig. 12.4 IMF5 to IMF8 of a male speaker

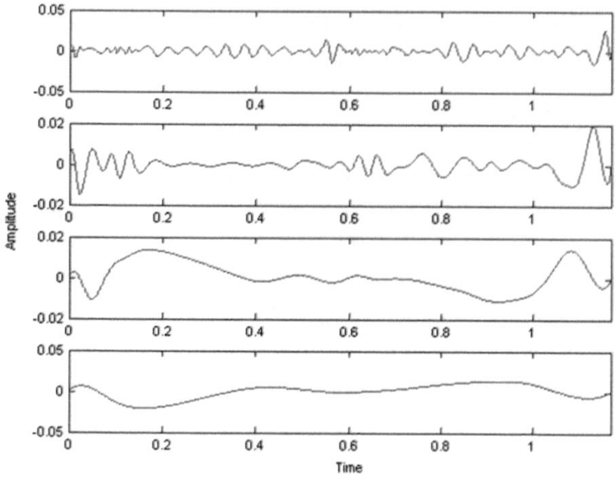

Fig. 12.5 IMF9 to IMF12 of a male speaker

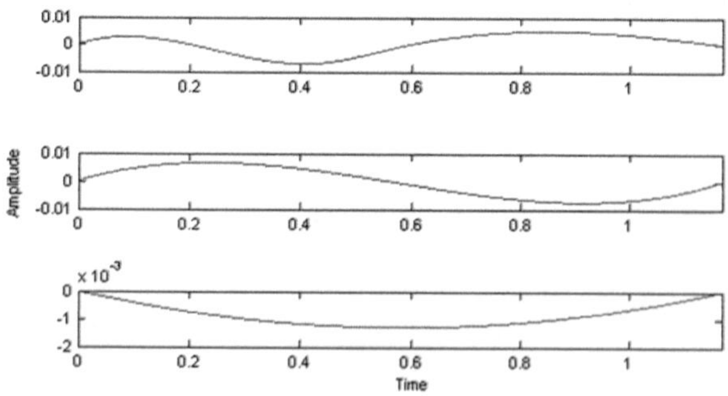

Fig. 12.6 IMF13, IMF14 and residual signal of a male speaker

speakers to observe the effectiveness of the reconstruction model. A similarity test between original and the reconstructed speech samples are done by finding correlation between the two as represented in Table 12.2 for the all the speakers

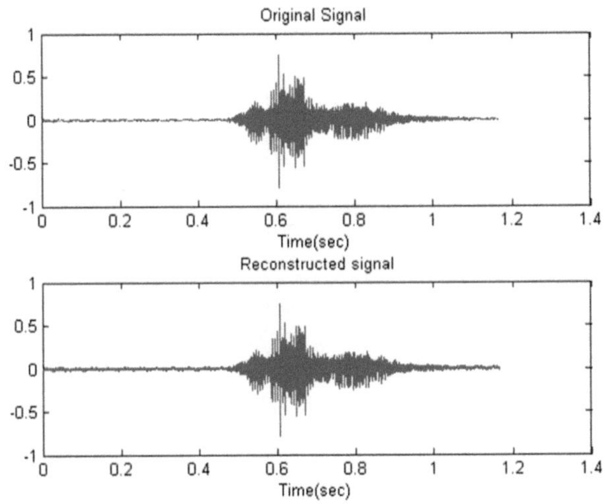

Fig. 12.7 Original and reconstructed speech signal of a male speaker

Table 12.2 Dissimilarity between original and reconstructed speech samples for male and female speech samples

Sl. no.	Speech sample	Dissimilarity measure (in %)
1	Female1	0.2
2	Female2	0.83
3	Female3	1.64
4	Female4	1.68
5	Female5	0.31
6	Female6	1.91
7	Female7	1.38
8	Female8	1.52
9	Female9	0.71
10	Female10	3.92
11	Male1	0.4
12	Male2	0.9
13	Male3	1.41
14	Male4	1.49
15	Male5	1.62
16	Male6	1.92
17	Male7	2.1
18	Male8	2.98
19	Male9	2.91
20	Male10	6.69

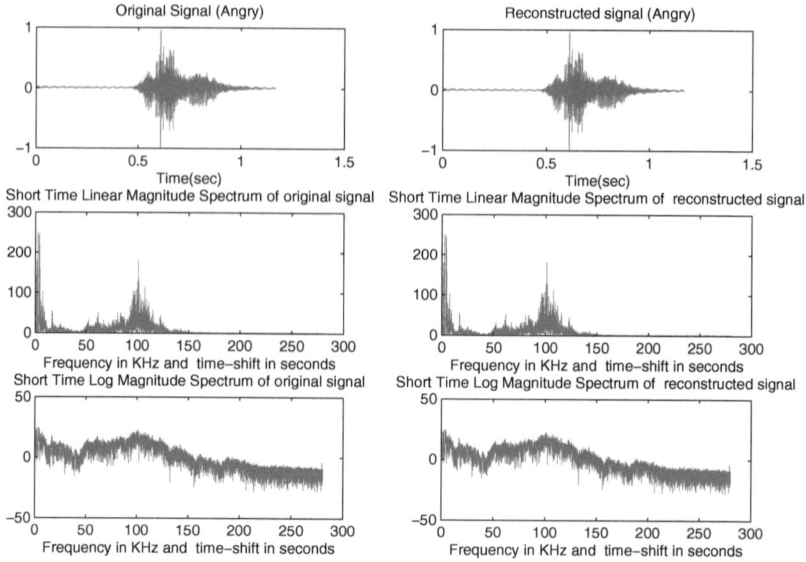

Fig. 12.8 Short-time linear and log-magnitude spectrum for original and reconstructed speech samples

used in the first round of experiment. Figure 12.8 represents the short-time linear magnitude spectrum and short-time log magnitude spectrum both for original and the reconstructed speech sample. The speech sample which we have taken in this case contain silent,voice and unvoiced parts.

2. **Proposed model with the variation of speakers mood**:
   The second round of experiment was carried out with speech samples recorded with speaker's mood variation. The speech samples are recorded in different moods like anger, loud, and soft, with the same set of speakers used in the first round and same recording specifications. EMD-based reconstruction algorithm is applied to this speech samples and the similarity measurement between original and reconstructed signal is performed. It is observed that the proposed algorithm gives satisfactory performance in this second round of experiment as well. Table 12.3 shows the dissimilarity measures for 10 such signals and Fig. 12.9 represents the original and reconstructed signal for three different moods.

3. **Proposed model in mobile channel data:**
   To check the effectiveness further, we have applied the proposed model in speech data collected during mobile telephonic conversation which has the sampling frequency of 8000 samples/s. The experimental results show the effectiveness of EMD-based approach for reconstruction of the speech signal in case mobile channel data. Table 12.4 shows the dissimilarity measures for 10 such signals

Table 12.3 Dissimilarity between original and reconstructed speech samples at three different mood (%)

Sl. no.	Loud mood	Angry mood	Normal mood
1	0.7	0.1	1.64
2	0.4	2.4	1.91
3	0.2	2.6	0.71
3	0.9	5.8	2.9
4	1	0.9	3.92
5	5.4	1.3	1.49
6	2.1	1.42	1.63
7	0.6	2.17	0.83
8	0.27	0.87	1.52
9	2.5	1.8	3.2

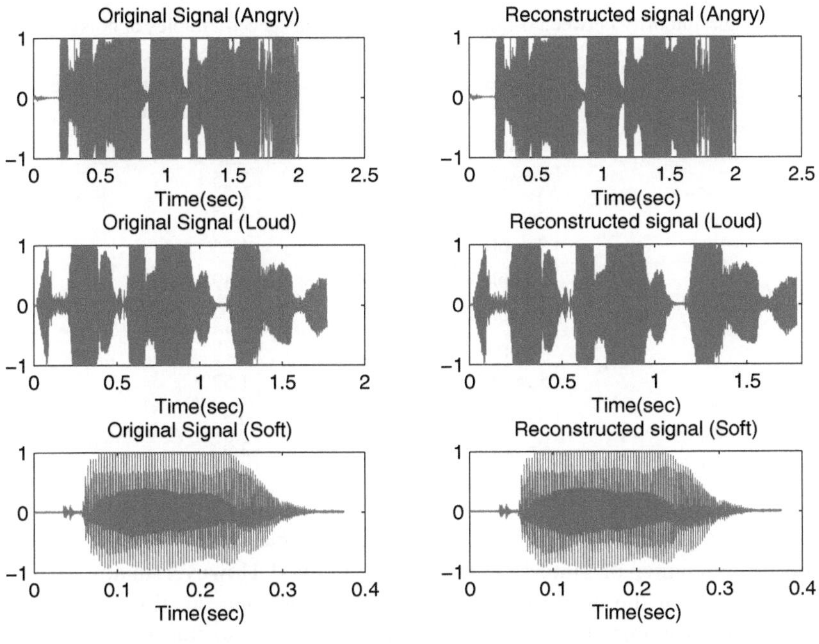

Fig. 12.9 Original and reconstructed speech signals at three different mood

4. **Proposed model in noisy speech signal:**
   Final round of experiment was performed on the noisy speech signal. We have used speech signals recorded under normal condition and added white Gaussian noise on it. If $x(t)$ is the original signal and $d(t)$ is white Gaussian noise then $n(t) = x(t) + d(t)$ is the signal that we have used as the input to the EMD block. Then the residual $r_n(t)$ is obtained which is the lowest frequency component of

Table 12.4 Dissimilarity between original and reconstructed speech signals collected in mobile channel (%)

Sl. no.	Speech sample	Mobile channel data
1	Sample1	2.3
2	Sample2	2.02
3	Sample3	2.51
4	Sample4	3.92
5	Sample5	2.22
6	Sample6	2.06
7	Sample7	0.17
8	Sample8	3.1
9	Sample9	6.6
10	Sample10	2.42

Table 12.5 Dissimilarity between original and reconstructed speech samples in noisy speech signals

Sl. no.	Speech sample	Dissimilarity measure (in percentage)
1	Sample1	0.2
2	Sample2	3.25
3	Sample3	1.65
4	Sample4	1.36
5	Sample5	0.49
6	Sample6	0.042
7	Sample7	1.38
8	Sample8	1.52
9	Sample9	6.3
10	Sample10	3.92

the noisy signal. If $r_n(t)$ is filtered with the original vocal tract response of the speech signal then original speech signal is obtained. However, in noisy condition the demerits of the proposed model is that the original vocal tract information should be available. Hence further study is required so that the model can be applied in blind situation. Table 12.5 represents the dissimilarity measures for 10 noisy signals.

12.4 Conclusion and Future Direction

Here we have discussed about a novel method for time-frequency analysis of speech signal which is used for glottal source information extraction. EMD is used for nonstationary and nonlinear signal processing. From experimental results, it can be seen that the EMD is a satisfactory method for decomposing speech signal into number of IMFs and obtaining the lowest frequency component, i.e., residual signal. This residual signal is considered as the source signal in case of a speech. Speech production model can be represented as source filter model. Applying EMD on different speech signals, source information have been found. Applying the concept of source filter model the vocal tract filter response is estimated. Later the original signal is reconstructed from the residue and vocal tract filter response, and finally original speech signal is reconstructed. The proposed approach next can be used for voice conversion application with the help of one person's source information and vocal tract filter of another person.

References

1. Bouzid A, Ellouze N EMD vocal tract frequency analysis of speech signal. In: 4th international conference sciences of electronic, technologies of information and telecommunications, PP TUNISIA
2. Sheth SS Extraction of pitch from speech signals using Hilbert Huang transform. http://www.iitg.ernet.in/scifac/cep/public_html/Resources/ece_kailash/files/07010245_ExtractionOfPitchFromSpeechSignalsUsingHilbertHuangTransform.pdf
3. Battista BM, Knapp C, McGee T, Goebel V (2007) Application of the empirical mode decomposition and Hilbert-Huang transform to seismic reflection data. Geophysics 72:29–37
4. Hasan T, Hasan K (2009) Suppression of residual noise from speech signals using empirical mode decomposition. IEEE Sign Process Lett 16(1):2–5
5. Schlotthauer G, Torres ME, Rufiner HL (2009) A new algorithm for instantaneous F0 speech extraction based on ensemble empirical mode decomposition. In: Proceedings of 17th European signal processing conference
6. Heydarian P, Reis JD (2005) Extraction of long-term structures in musical signals using the empirical mode decomposition. In: Proceedings of the 8th international conference on digital audio effects, Madrid, Spain
7. Schlotthauer G, Torres ME, Rufiner HL (2009) Voice fundamental frequency extraction algorithm based on ensemble empirical mode decomposition and entropies. In: Proceedings of the world congress on medical physics and biomedical engineering, Germany
8. Douglas O (2000) Speech communication human and machine, 2nd edn. IEEE Press, New York

Chapter 13
Assamese Vowel Speech Recognition Using GMM and ANN Approaches

Debashis Dev Misra, Krishna Dutta, Utpal Bhattacharjee, Kandarpa Kumar Sarma and Pradyut Kumar Goswami

Abstract This work focuses on the classification of Assamese vowel speech and recognition using Gaussian mixture model (GMM). The results are compared to the results obtained using artificial neural network (ANN). The training data is composed of a database of eight different vowels of Assamese language with 10 different recorded speech samples of each vowel as a set in noise-free and noisy environments. The testing data similarly is composed of the same number of vowels with each vowel containing 23 different recorded samples. Cepstral mean normalization (CMN) and maximum likelihood linear regression (MLLR) are used for speech enhancement of the data which is degraded due to noise. Feature extraction is done using mel frequency cepstral coefficients (MFCC). GMM and ANN approaches are used as classifiers for an automatic speech recognition (ASR) system. We found the success rate of the GMM to be around 81 % and that of the ANN to be above 85 %.

Keywords Cepstral mean normalization (CMN) · Maximum likelihood linear regression (MLLR) · Vowel · Gaussian mixture model (GMM) · Artificial neural network (ANN)

D.D. Misra (✉)
Royal School of Engineering and Technology, Guwahati, Assam, India
e-mail: debashish.dm@gmail.com

K. Dutta
NIT Nagaland, Dimapur, Nagaland, India
e-mail: krishnadutta54@gmail.com

U. Bhattacharjee
Department of Computer Science and Engineering, Rajiv Gandhi University, Doimukh, Arunachal Pradesh, India
e-mail: utpal.bhattacharjee@rgu.ac.in

K.K. Sarma
Department of Electronics and Communication Technology, Gauhati University, Guwahati 781014, Assam, India
e-mail: kandarpaks@gmail.com

P.K. Goswami
Assam Science and Technology University, Guwahati 781014, Assam, India
e-mail: pradyutgoswami@yahoo.com

K.K. Sarma et al. (eds.), *Recent Trends in Intelligent and Emerging Systems*, Signals and Communication Technology, DOI 10.1007/978-81-322-2407-5_13

13.1 Introduction

Vowels are voiced sound during the production of which sound obstruction occurs in the oral or nasal cavities. Voiced speech is a sound produced with the vibration of vocal cords. In the speech, vowels are produced by exciting an essentially fixed vocal tract, shaped with quasiperiodic pulses of air caused by the vibration of the vocal cords [1].

Assamese is an eastern Indo-Aryan language spoken by about 20 million people in the Indian states of Assam, Meghalaya, and Arunachal Pradesh, and also spoken in Bangladesh and Bhutan [2]. There are 11 vowels in Assamese language and are distinguished by the place of articulation (front, central or back) and the position of the tongue (high, mid or low). The way in which the cross-sectional area varies along the vocal tract determines the resonance frequencies of the tract (the formants) and thereby the sound that is produced. The vowel sound produced is determined primarily by the position of the tongue, but the position of the jaw, lips, and to a small extent, the velum also influence the resulting sound [1].

This work focuses on the classification of Assamese vowel speech and recognition using gaussian mixture model (GMM). The results are compared to the results obtained using artificial neural network (ANN). The training data is composed of a database of 8 different vowels of Assamese language with 10 different recorded speech samples of each vowel as a set in noise-free and noisy environments. The testing data similarly is composed of the same number of vowels with each vowel containing 23 different recorded samples. Cepstral mean normalization (CMN) and maximum likelihood linear regression (MLLR) are used for speech enhancement of the data which is degraded due to noise. Feature extraction is done using mel frequency cepstral coefficients (MFCC). GMM and ANN approaches are used as classifiers for an automatic speech recognition (ASR) system. We found the success rate of the GMM to be around 81 % and that of the ANN to be above 85 %. Some of the related literature are [1, 3–8].

The rest of the paper is organized as follows: in Sect. 13.2, we briefly discuss about the basic notions related to the work. The system model is described in Sect. 13.3. The experimental details and results are discussed in Sect. 13.4. The work is concluded in Sect. 13.5.

13.2 Theoretical Considerations

Here, a brief discussion about ANN and GMM is given.

13.2.1 ANN

ANN: ANNs are bio-inspired computational tools that provide human-like performance in the field of ASR. These models are composed of many nonlinear computational elements called perceptrons operating parallel in patterns similar to the biological neural networks [9]. ANN has been used extensively in ASR field during the past two decades. The most beneficial characteristics of ANNs for solving ASR problem are the fault tolerance and nonlinear property. The earliest attempts involved highly simplified tasks, e.g., classifying speech segments as voiced/unvoiced or nasal/fricative/plosive. Success in these experiments encouraged researchers to move on to phoneme classification. The basic approaches to speech classification using ANN are static and dynamic.

In static classification, the ANN accepts the input speech and makes a single decision. By contrast, in dynamic classification, the ANN considers only a small window of the speech. This window slides over the input speech while the ANN generates decisions. Static classification works well for phoneme recognition, but it scales poorly to the level of words or sentences. In contrast, dynamic classification scales better. Either approach may make use of recurrent connections, although recurrence is more often found in the dynamic approach [4, 5].

13.2.2 GMM

The GMM is a density estimator and is one of the most commonly used types of classifier. In this method, the distribution of the feature vector x is modeled clearly using a mixture of M Gaussians. A GMM is modeled by many different Gaussian distributions. Each of the Gaussian distribution has its mean, variance, and weights in the GMM. A Gaussian mixture density is a weighted sum of M component densities (Gaussians) as depicted in following figure and given by equation.

$$p(\overrightarrow{x}|\lambda) = \sum_{i=1}^{M} p_i x_i(\overrightarrow{x}) \tag{13.1}$$

where x is a L dimensional vector, p_i are mixture weights, and $b_i(x)$ are component densities with $i = 1M$. Each component density is a L variate Gaussian function of the form,

$$b_i(x) = \frac{1}{(2\pi)^{L/2}|\sum i|^{1/2}} \exp\left(-\frac{1}{2}(x-\mu_i)' \sum_{i=1}^{-1}(x-\mu_i)\right) \tag{13.2}$$

where μ_i is the mean and $\sum i$ is covariance matrix. The mixture weights satisfy the constraint that $\sum_{i=1}^{M} p_i$. T is the total number of feature vectors or total number of frames. T is the total number of feature vectors or total number of frames. The mean vectors, covariance matrices, and mixture weights of all Gaussians together represent a speaker model and parameterize the complete Gaussian mixture density. GMMs are commonly used as a parametric model of the probability distribution of continuous measurements or features in a biometric system, such as vocal tract-related

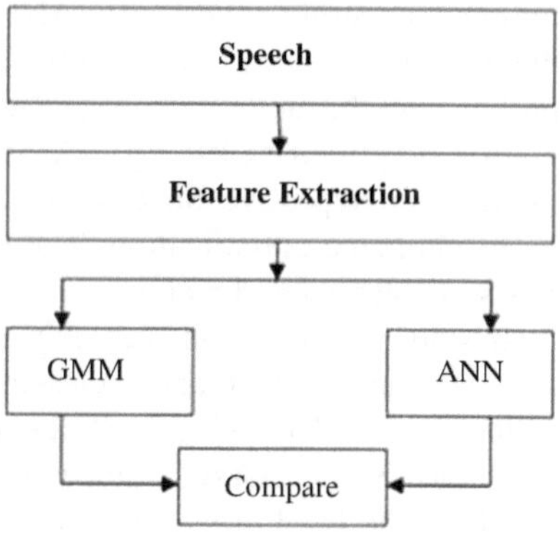

Fig. 13.1 System model

spectral features in a speaker recognition system. GMM parameters are estimated from training data using the iterative expectation-maximization (EM) algorithm or maximum a posteriori (MAP) estimation from a well-trained prior model [6, 7]. GMMs are often used in biometric systems, most notably in speaker recognition systems, due to their capability of representing a large class of sample distributions. One of the powerful attributes of the GMM is its ability to form smooth approximations to arbitrarily shaped densities.

13.3 System Model

In this study, we concentrate on Assamese vowel speech recognition using GMM and compare the results obtained using ANN. CMN and MLLR are used for speech enhancement of the data degraded due to noise. Feature extraction is done using MFCC. GMM and ANN approaches are used as classifiers for an ASR system for Assamese speech.

The system model is shown in Fig. 13.1. Feature extraction is the estimation of variables (feature vector) from the observation of a speech signal which contains different information such as dialect, context, speaking style, and speaker emotion. It estimates a set of features from the speech signal that represents some speaker-specific information. The aim is to transform the speech signal into a collection of variables that can preserve the signal information and that can be used to make comparisons [8].

13.4 Experimental Details and Results

Initially, we record certain number of vowel speech samples of Assamese language out of which some are retained in a clean form and a few are corrupted. In the proposed system, data is collected in 16 kHz sampling rate at 16 b mono format. Speech data collected is grouped into frame of 30 ms with one-third overlapping. It gives a frame rate of 10 ms. After pre-emphasis, each frame is multiplied by Mel-filter bank with 20 filters and the MFCC coefficients are calculated. Here, 19 MFCC coefficients are considered along with their first-order derivatives as feature vector for each frame. Thus, we extract features using MFCC. The feature set contains samples which are clean and noise corrupted.

These are next modeled using GMM and applied to ANN for training. There is a training phase during which the GMM and ANN learns. The samples sets have the clean and noise-corrupted sets. Next, test and validation processes are performed during which the GMM and ANN demonstrate the decision-making role as part of the ASR. The speech samples derived from the inputs before feeding to GMM and ANN are enhanced by the CMN and MLLR approaches which contribute to the performance of the system. CMN has been done for cepstral coefficients extracted from the speech signal. After CMN, the model-based algorithm MLLR has been used for further noise elimination. MLLR is a model-based compensation method. It uses a mathematical model of the environment and attempts to use samples of the degraded speech to estimate the parameters of the model. In order to evaluate the clean speech in a real environment condition, the clean speech is deteriorated by adding the white Gaussian noise. The assumptions made are that the noise is additive and not correlated with the speech signal.

The noisy signal due to the addition of the white Gaussian noise is shown in the second plot of the below Fig. 13.4. The signal-to-noise ratio (SNR) is set at 5 dB. After applying CMN and MLLR to the noisy signal, the enhanced speech signal is shown below in Fig. 13.2. The original speech signal is plotted in blue, the signal checked for enframing and deframing is shown in black and the average noised removed signal using CMN and MLLR is shown in red.

A similar set of results are generated using a combination of CMN and MLLR shown in Fig. 13.3.

The training data is composed of a database of eight different Assamese vowels with 10 different recoded speech samples of each vowel as a set in noisy environments. The testing data similarly was composed of the same number of vowels with each vowel containing 23 different recorded samples (Fig. 13.4).

We have here used a recurrent neural network (RNN) of two hidden layers trained with error backpropagation through time (BPTT) algorithm [9]. The RNN is a special form of ANN with the ability to track time variations in input signals. The

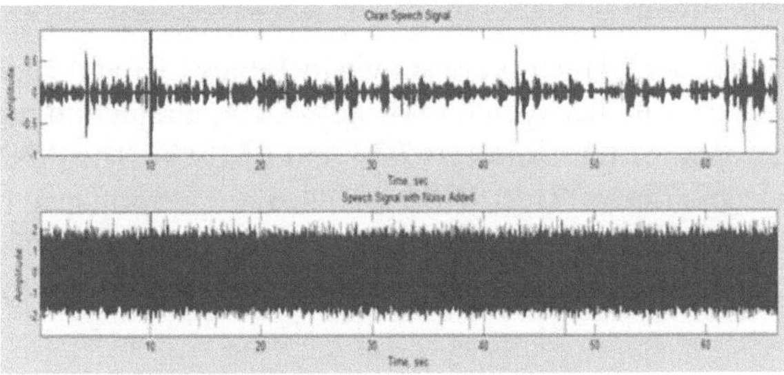

Fig. 13.2 Clean speech signal and speech signal with white Gaussian noise added

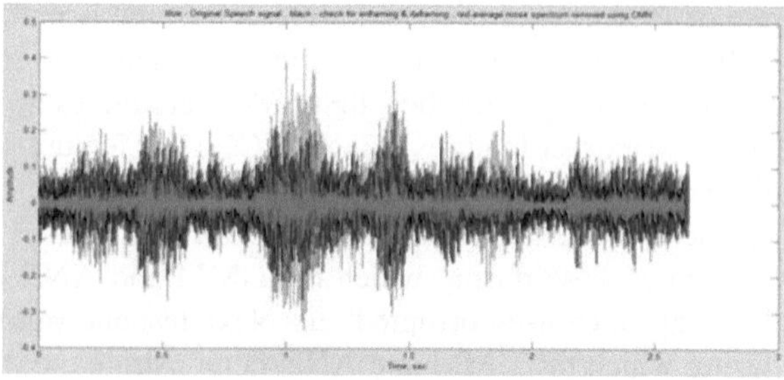

Fig. 13.3 Clean signal in *blue*, check signal in *black*, and enhanced signal in *red*

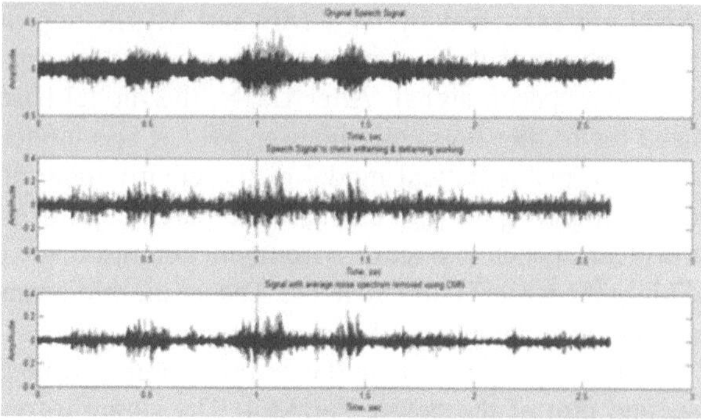

Fig. 13.4 Clean, check, and enhanced signal derived using CMN and MLLR

Table 13.1 Results derived using ANN

Sl. no.	Input class	Recognition rate (%)	False rejection rate (%)	False acceptance rate (%)
1	Class 1	87.60	10.04	2.35
2	Class 2	83.26	10.04	6.69
3	Class 3	91.95	5.69	2.34
4	Class 4	78.91	10.04	11.04
5	Class 5	91.95	1.34	4.69
6	Class 6	83.26	5.69	11.04
7	Class 7	91.95	5.69	2.34
8	Class 8	71.21	18.73	11.04
Overall		85.01	9.83	5.44

experimental results are shown in Tables 13.1 and 13.2. The GMM approach shows a success rate between 65.86 and 87.6 %, a rejection rate of 5.69–23.08 %, and a false acceptance rate between 2.35 and 15.39 %. The ANN, on the other hand, shows a success rate of 71.2–91.95 %, rejection performance between 1.34 and 18.73 %, and a false acceptance rate of 2.34–11.04 %. This improved performance of the ANN is due to its robustness, adaptive learning, and ability to retain the learning.

Table 13.2 Results derived using GMM

Sl. no.	Input class	Recognition rate (%)	False rejection rate (%)	False acceptance rate (%)
1	Class 1	83.26	14.39	2.35
2	Class 2	78.91	15.39	5.69
3	Class 3	87.60	10.04	2.35
4	Class 4	74.56	10.04	15.39
5	Class 5	87.60	5.69	5.69
6	Class 6	78.91	10.04	11.04
7	Class 7	91.95	5.69	2.35
8	Class 8	65.86	23.08	11.04
Overall		81.08	11.80	6.98

Table 13.3 Results showing computational complexity in GMM and ANN

Algorithm	GMM	ANN
Time (s)	64.29	50.45

The computational requirements of the two approaches recorded during training is shown in Table 13.3. It shows that the ANN takes lesser time to complete the processing. Thus, in terms of higher recognition accuracy, lower rejection, and false acceptance rates and reduced computational requirement, the ANN-based approach is superior compared to the GMM approach.

13.5 Conclusion

This work focuses on the classification of Assamese vowel speech and recognition using GMM and ANN. CMN and MLLR are used for speech enhancement of the data which is degraded due to noise. Feature extraction is done using MFCC. GMM and ANN approaches are used as classifiers for an ASR system. We found success rate of the GMM to be around 81 % and that of ANN to be above 85 %. The GMM approach shows a success rate between 65.86 to 87.6 %, a rejection rate of 5.69 to 23.08 % and a false acceptance rate between 2.35 to 15.39 %. The ANN, on the other hand, shows a success rate of 71.2 to 91.95 %, rejection performance between 1.34 to 18.73 % and a false acceptance rate of 2.34 to 11.04 %. The ANN further takes atleast 21 % lower computational time compared to the GMM approach. This improved performance of the ANN is due to its robustness, adaptive learning and ability to retain the learning.

References

1. Sarma M, Sarma KK (2013) Vowel phoneme segmentation for speaker identification using an ANN-based framework. J Intell Syst 22(2):111–130
2. Gait EA (1905) A history of assam (1926), Rev. edn. Thacker Spink and Co, Calcutta
3. Zhao B, Schultz T (2002) Towards robust parametric trajectory segmental model for vowel recognition. In: Proceedings of the IEEE international conference on acoustics, speech, and signal processing
4. Lippmann RP (1989) Review of neural networks for speech recognition. Neural Comput Spring 1(1):1–38
5. Huang WY, Lippmann RP (1987) Neural net and traditional classifiers. In: Proceedings of IEEE conference on neural information processing systems—natural and synthetic, New York
6. Reynolds DA, Rose RC (1995) Robust text-independent speaker identification using Gaussian mixture speaker models. IEEE Trans, Speech Audio Process, p 3
7. Hasan R, Jamil M, Rabbani G, Rahman S (2004) Speaker identification using mel frequency cepstral coefficients. In: Proceedings of 3rd international conference on electrical and computer engineering (ICECE 2004), Dhaka, Bangladesh
8. Sarma M, Sarma KK (2013) An ANN based approach to recognize initial phonemes of spoken words of assamese language. Appl Soft Comput 13(5):2281–2291
9. Haykin S (2009) Neural network and learning machine, 3rd edn. PHI Learning Private Limited, New Delhi

Part V
Review Chapters on Selected Areas

This is the last part of the volume with three review works related to important segments, namely speech recognition, VLSI design, and communication. M. Sarma et al. provides an exhaustive account of the works related to speech recognition in Indian languages. The work is expected to provide certain insights into the area and help upcoming researchers contemplating works in such domains. The next work is a review by M.P. Sarma, which highlights some recent trends in VLSI designs related to communication applications. The concluding chapter of the volume is another review which highlights the use of soft computing tools in wireless communication.

Chapter 14
Speech Recognition in Indian Languages—A Survey

Mousmita Sarma and Kandarpa Kumar Sarma

Abstract In this paper, a brief overview derived out of detailed survey of speech recognition works reported in Indian languages is described. Robustness of speech recognition systems toward language variation is the recent trend of research in speech recognition technology. To develop a system which can communicate with human in any language like any other human is the foremost requirement in order to design appropriate speech recognition technology for one to all. India is a country which has vast linguistic variations among its billion plus population. Therefore, it provides a sound area of research toward language-specific speech recognition technology. From the beginning of the commercial availability of the speech recognition system, the technology has been dominated by the hidden Markov model (HMM) methodology due to its capability of modeling temporal structures of speech and encoding them as a sequence of spectral vectors. Most of the work done in Indian languages also uses HMM technology. However, from the last 10–15 years after the acceptance of neurocomputing as an alternative to HMM, artificial neural network (ANN)-based methodologies have started to receive attention for application in speech recognition. This is a trend worldwide as part of which few works have also been reported by a few researchers.

Keywords Speech recognition system · Indian languages · Hidden Markov model (HMM) · Gaussian mixture model (GMM) · Artificial neural network (ANN)

M. Sarma (✉)
Department of Electronics and Communication Engineering,
Gauhati University, Guwahati, Assam, India
e-mail: go4mou@gmail.com

K.K. Sarma
Department of Electronics and Communication Technology, Gauhati University,
Guwahati 781014, Assam, India
e-mail: kandarpaks@gmail.com

K.K. Sarma et al. (eds.), *Recent Trends in Intelligent and Emerging Systems*,
Signals and Communication Technology, DOI 10.1007/978-81-322-2407-5_14

14.1 Introduction

The problem of automatic speech recognition (ASR) was at the forefront of research till 1930 when the first electronic voice synthesizer was designed by Homer Dudley of Bell Laboratories. After that, ASR lost its fascination among the speech processing community. Probably that was the starting of research in the direction of designing a machine that can mimic the human capability of speaking naturally and responding to spoken languages. Initial developments cover a simple machine that responds to isolated sounds. Recently, speech recognition technology has risen to such a height that a large community of people now talk to their smartphones, asking them to send e-mail and text messages, search for directions, or find information on the web. However, speech recognition technology is still far from having a machine that converses with humans on any topic like another human. In the present time, research in speech recognition concentrates on developing systems that can show robustness for variability in environment, speaker, and language. India is a linguistically rich country having 22 official languages and hundreds of other sublanguages and dialects spoken by various communities covering the billion plus population. Communication among human beings is dominated by spoken language. Therefore, it is natural for people to expect speech interfaces with computers which can speak and recognize speech in native language. But speech recognition technology in the Indian scenario is restricted to small amount of people who are both computer literate and proficient in written and spoken English. In this domain, extensive researches are going on all over India among various groups to make appropriate ASR systems in Indian languages.

Initial speech recognition systems were on isolated word recognition designed to perform special task. But in the last 25 years, certain dramatic progress in statistical methods for recognizing speech signals has been noticed. The statistical approach makes use of the four basic principles which are Bayes decision rule for minimum error rate, probabilistic models, e.g., hidden Markov models (HMMs), or conditional random fields (CRF) for handling strings of observations like acoustic vectors for ASR and written words for language translation, training criteria, and algorithms for estimating the free-model parameters from large amounts of data and the generation or search process that generates the recognition or translation result [1, 2]. The speech recognition research is dominated by the statistical approaches specifically by the HMM technology till the last one decade. It is the improvement provided by HMM technology for speech recognition in the late 1970s and the simultaneous improvement in speed of computer technology, due to which the ASR systems have become commercially viable in 1990s [3–5]. But recently in the last decade, all over the world, the ANN-based technologies are gaining attention. This is due to the fact that ANN models are composed of many nonlinear computational elements operating in parallel and arranged in the pattern of biological neural network. It is expected that human neural network like models may ultimately be able to solve the complexities of speech recognition system and provide human-like performance.

In this article, we have highlighted some works related to the research and development of ASR in Indian languages during the last decade so as to provide a picture of the fundamental progress that has been made in the large variety of Indian languages. The survey is divided into two groups based on statistical- and ANN-based technology. Initially, a glance of early speech recognition technology in world languages is also included.

14.2 Early Speech Recognition Technology

Early speech recognition systems used the acoustic-phonetic theories of speech to determine the feature [1]. Due to the complexity of human language, the inventors and engineers first focused on number or digit recognition. The first speech recognition system was built in Bell Laboratories by Davis, Biddulph, and Balashek in 1952 which could understand only isolated digits for a single speaker [6]. They used the formant frequency measured during vowel regions of each digit as a feature. During 1950–1970, laboratories in the United States, Japan, England, and the Soviet Union developed other hardware dedicated to recognizing spoken sounds, expanding speech recognition technology to support four vowels and nine consonants [7–13]. In the 1960s, several Japanese laboratories demonstrated their capability of building special-purpose hardware to perform a speech recognition task. Among them, the vowel recognizer of Suzuki and Nakata at the Radio Research Lab in Tokyo [7], the phoneme recognizer of Sakai and Doshita at Kyoto University [8], and the digit recognizer of NEC Laboratories [9] were most notable. The work of Sakai and Doshita involved the first use of a speech segmenter for analysis and recognition of speech in different portions of the input utterance. An alternative to the use of a speech segmenter was the concept of adopting a nonuniform time scale for aligning speech patterns [11, 12], dynamic programming for time alignment between two utterances known as dynamic time warping, in speech pattern matching [12] etc. Another milestone of 1960s is the formulation of fundamental concepts of linear predictive coding (LPC) [14, 15] by Atal and Itakura, which greatly simplified the estimation of the vocal tract response from speech waveforms. Development during 1970s includes the first speech recognition commercial company called Threshold Technology founded by Martin [1] and speech understanding research (SUR) program founded by advanced research projects agency (ARPA) of the U.S. Department of Defense [1]. Threshold Technology later developed the first real ASR product called the VIP-100 System [1] for some limited application and Carnegie Mellon University under ARPA developed Harpy system which was able to recognize speech using a vocabulary of 1011 words with reasonable accuracy. The Harpy system was the first to take advantage of a finite-state network to reduce computation and efficiently determine the closest matching string [16]. DRAGON system by Jim Baker was also developed during 1970s [17]. In 1980s, speech recognition turned toward prediction. Speech recognition vocabulary improved from about a few hundred words to several thousand words and had the potential to recognize an unlimited number of words.

The major reason for this up gradation is the new statistical method HMM. Rather than simply using templates for words and looking for sound patterns, HMM considered the probability of unknown sounds being words. The foundations of modern HMM-based continuous speech recognition technology were laid down in the 1970s by groups at Carnegie Mellon and IBM who introduced the use of discrete density HMMs [16–18], and then later at Bell Laboratories [19–21] where continuous density HMMs were introduced. Another reason of this drastic improvement of the speech recognition technology is the application of fundamental pattern recognition technology to speech recognition based on LPC methods in the mid 1970s by Itakura [22], Rabiner and Levinson [23], and others. Due to the expanded vocabulary provided by HMM methodology and the computer with faster processor, in 1990s speech recognition software become commercially available.

During 1980s, ANN technology was also introduced in the domain of speech recognition. The brains impressive superiority at a wide range of cognitive skills like speech recognition, has motivated the researchers to explore the possibilities of ANN models in the field of speech recognition in 1980s [24], with a hope that human neural network like models may ultimately lead to human-like performance. Early attempts at using neural networks for speech recognition centered on simple tasks like recognizing a few phonemes or a few words or isolated digits, with good success [25–27], using pattern mapping by multilayer perceptron (MLP). But at the later half of 1990, suddenly ANN-based speech research got terminated [24] after the statistical framework HMM come into focus, which supports both acoustic and temporal modeling of speech. However, it should be mentioned that the current best systems are far from equaling human-like performance and many important research issues are still to be explored. Therefore, the value of ANN-based research is still large and nowadays it is considered as the hot field in the domain of speech recognition.

14.3 Research of Speech Recognition in Indian Languages

The current speech researchers are focused on using technology to overcome the challenges in natural language processing, so that next-generation speech recognition system can provide easy and natural modes of interaction for its customer. Specifically, it has become the primary concern for the scientist and engineers to build systems that can be consumed by the common public and to enable natural language transactions between human and machine. The language-specific speech recognition is difficult mainly because the system requires knowledge of word meaning, communication context, and the commonsense. This variability includes the effect of the phonetic, phonology, syntax, semantic, and communication modes of the speech signal. While having the different meaning and usage patterns, words can have the same phonetic realization. If the words were always produced in the same way, speech recognition would be relatively easy. However, for various reasons words are almost always pronounced differently due to which it is still a challenge to build a recognizer that is robust enough in case of any speaker, any language, and any speaking environment.

India is the country where vast cultural and linguistic variations are observed. Therefore, in such a multilingual environment there is a huge possibility of implementing speech technology. The constitution of India, has recognized 17 regional languages (Assamese, Bengali, Bodo, Dogri, Gujarati, Kannada, Kashmiri, Konkani, Maithili, Malayalam, Manipuri, Marathi, Nepali, Oriya, Punjabi, Sanskrit, Santhali, Sindhi, Tamil, Telugu, Urdu) along with Hindi which is the national language of India. However, till date the amount of work done in speech recognition in Indian languages has not reached the domain of rural and computer illiterate people of India [28]. Few attempts have been made by HP Labs India, IBM research lab, and some other research groups. Yet there is lots of scopes and possibilities to be explored to develop speech recognition system using Indian languages.

After the commercial availability of speech recognition system, the HMM technology has dominated the speech research. HMMs lie at the heart of virtually all modern speech recognition systems. The basic HMM framework has not changed significantly in the last decades, but various modeling techniques have been developed within this framework that has made the HMM technology considerably sophisticated [14, 15, 22]. At the same time, from the last one or two decades, ANN technology has also been used by various researchers. The current state has considered that HMM has given the best it could, but in order to improve the accuracy of speech recognition technology under language, speaker, and environmental variations, other technology is required. In the Indian language scenario, the speech recognition work can be reviewed in two parts: work done using statistical framework like HMM, gaussian mixture models (GMM) and a very few work done using ANN technology. Further a few hybrid technology-based work is also found in the literature. The following sections describe the speech recognition work developed in Indian languages over the last decade.

14.3.1 Statistical Approach

The basic statistical method used in speech recognition purpose is the HMM methodology. HMMs are a parametric model which can be used to model any time series but particularly suitable to model speech event. In HMM-based speech recognition, it is assumed that the sequence of observed speech vectors corresponding to each word is generated by a Markov model [29]. The forward backward reestimation theory called the Baum–Welch reestimation used in HMM-based speech recognition modifies the parameter in every iteration and the probability of training data increases until a local maxima reach. The success of HMM technology lies on its capability to estimate a extended set of unknown utterance from a known set of utterance given as training set [2, 30, 31]. The availability of well-structured software like hidden Markov model tool kit (HTK) [30] and CMUs Sphinx [31] which can successfully implement HMM technology makes it easier for further research and development to incorporate new concepts and algorithms in speech recognition.

A few relevant work done in Indian languages using statistical framework like HMM are discussed below.

1. In the journal Sadhana in 1998, a work [32] has been reported by Samudravijaya et al., where they have presented a description of a speech recognition system for Hindi. The system follows a hierarchic approach to speech recognition and integrates multiple knowledge sources within statistical pattern recognition paradigms at various stages of signal decoding. Rather than making hard decisions at the level of each processing unit, relative confidence scores of individual units are propagated to higher levels. A semi-Markov model processes the frame level outputs of a broad acoustic maximum likelihood classifier to yield a sequence of segments with broad acoustic labels. The phonemic identities of selected classes of segments are decoded by class-dependent ANNs which are trained with class-specific feature vectors as input. Lexical access is achieved by string matching using a dynamic programming technique. A novel language processor disambiguates between multiple choices given by the acoustic recognizer to recognize the spoken sentence. The database used for this work consisted of sentences having 200 words, which are most commonly used in railway reservation enquiry task.
2. Another work [33] by Rajput et al. from IBM India Research Lab has been reported in 2000, where they have attempted to build decision trees for modeling phonetic context dependency in Hindi by modifying a decision tree built to model context dependency in American English. In a continuous speech recognition system, it is important to model the context-dependent variations in the pronunciations of phones. Linguistic-Phonetic knowledge of Hindi is used to modify the English phone set. Since the Hindi phone set being used is derived from the English phone set, the adaptation of the English tree to Hindi follows naturally. The method may be applicable for adapting between any two languages.
3. In 2008, Kumar et al. of IBM India Research Lab developed another HMM-based large vocabulary continuous speech recognition system for Hindi language. In this work [34], they have presented two new techniques that have been used to build the system. Initially, a technique for fast bootstrapping of initial phone models of a new language is given. The training data for the new language is aligned using an existing speech recognition engine for another language. This aligned data is used to obtain the initial acoustic models for the phones of the new language. Following this approach requires less training data. They have also presented a technique for generating baseforms, i.e., phonetic spellings for phonetic languages such as Hindi. As is inherent in phonetic languages, rules generally capture the mapping of spelling to phonemes very well. However, deep linguistic knowledge is required to write all possible rules, and there are some ambiguities in the language that are difficult to capture with rules. On the other hand, pure statistical techniques for baseform generation require large

amounts of training data, which are not readily available. But here they have proposed a hybrid approach that combines rule-based and statistical approaches in a two-step fashion.

4. For Hindi language, Gaurav et al. has reported another work recently in 2012 [35]. A continuous speech recognition system in Hindi is tailored to aid teaching geometry in primary schools. They have used the mel frequency cepstral coefficients (MFCC) as speech feature parameters and HMM to model these acoustic features. The Julius recognizer which is language independent was used for decoding.
5. Kumar et al. in 2012 has designed a feature extraction modules ensemble of MFCC, LPCC, perceptual linear predictive analysis (PLP) etc. to improve Hindi speech recognition system [36]. The outputs of the ensemble feature extraction modules have been combined using voting technique ROVER.
6. Bhuvanagirir and Kopparapu have reported another work on mixed language speech recognition [37] Hindi and English combination in 2012.
7. In 2008 Thangarajan et al. has reported a work in continuous speech recognition for Tamil Language [38]. They have built a HMM-based continuous speech recognizer based on word and triphone acoustic models. In this experiment, a word-based context-independent (CI) acoustic model for 371 unique words and a triphone-based context-dependent (CD) acoustic model for 1700 unique words have been built for Tamil language. In addition to the acoustic models, a pronunciation dictionary with 44 base phones and trigram-based statistical language model have also been built as integral components of the linguist. These recognizers give satisfactory word accuracy for trained and test sentences read by trained and new speakers.
8. In 2009, Kalyani and Sunithato worked toward the development of a dictation system like Dragon for Indian languages. In their paper [39], they have focused on the importance of creating speech database at syllable units and identifying minimum text to be considered while training any speech recognition system. They have also provided the statistical details of syllables in Telugu and its use in minimizing the search space during recognition of speech. The minimum words that cover maximum syllables are identified which can be used for preparing a small text for collecting speech sample while training the dictation system.
9. Another work in Telugu language is reported by Usha Rani and Girija in 2012. To improve the speech recognition accuracy on Telugu language, they have explored means to reduce the number of the confusion pairs by modifying the dictionary, which is used in the decoder of the speech recognition system. In their paper [40], they have described the different types of errors obtained from the decoder of the speech recognition system.
10. Das et al. reported a work [41] in Bengali language, where they have described the design of a speech corpus for continuous speech recognition. They have developed speech corpora in phone and triphone labeled between two age groups—20 to 40 and 60 to 80. HMM is used to align the speech data statistically and observed good performance in phoneme recognition and continuous word recognition done using HTK and SPHINX.

11. In Punjabi language, Dua, Aggarwal, Kadyan and Dua have reported a work in 2012, where they have attempted to develop a isolated word recognition system using HTK [42].
12. In 2013, another work has been reported by Mehta et al. where a comparative study of MFCC and LPC for Marathi isolated word recognition system is described [43].
13. Udhyakumar et al. reported a work [44] in 2004 for multilingual speech recognition to be used for information retrieval in Indian context. This paper analyzes various issues in building a HMM-based multilingual speech recognizer for Indian languages. The system is designed for Hindi and Tamil languages and adapted to incorporate Indian accented English. Language-specific characteristics in speech recognition framework are highlighted. The recognizer is embedded in information retrieval applications and hence several issues like handling spontaneous telephony speech in real-time, integrated language identification for interactive response and automatic grapheme to phoneme conversion to handle out of vocabulary words are addressed in this paper.
14. Some issues about the development of speech databases of Tamil, Telugu and Marathi for large vocabulary speech recognition system is reported in a work [45] by Anumanchipalli et al. in 2005. They have collected speech data from about 560 speakers in these three languages. They have also presented the preliminary speech recognition results using the acoustic models created on these databases using Sphinx 2 speech tool kit, which shows satisfactory improvement in accuracy.
15. During 2009, Hema A Murthy et al. described in [46, 47], a novel approach to build syllable-based continuous speech recognizers for Indian languages, where a syllable-based lexicon and language model are used to extract the word output from the HMM-based recognizer. The importance of syllables as the basic sub-word unit for recognition has been a topic for research. They have shown that a syllabified lexicon helps hypothesize the possible set of words, where the sentence is constructed with the help of N-gram based statistical language models. The database used for these works is the Doordarshan database, which is made of news bulletins of approximately 20 min duration of both male and female speakers.
16. Bhaskar et al. reported another work [48] on multilingual speech recognition in 2012. They have used a different approach to multilingual speech recognition, in which the phone sets are entirely distinct but the model has parameters not tied to specific states that are shared across languages. They have tried to build a speech recognition system using HTK for Telugu language. The system is trained for continuous Telugu speech recorded from native speakers.
17. In 2013, a work [49] has been reported by Mohan et al. where a spoken dialogue system is designed to use in agricultural commodities task domain using real-world speech data collected from two linguistically similar languages of India Hindi and Marathi. They have trained a subspace gaussian mixture model (SGMM) under a multilingual scenario [50, 51]. To remove acoustic, channel, and environmental mismatch between datasets from multiple languages, they

have used a cross-corpus acoustic normalization procedure which is a simple variant of speaker adaptive training (SAT) described by Mohan et al. in 2012 [52]. The resulting multilingual system provides the best speech recognition performance of 77.77 % for both languages .

14.3.2 Artificial Neural Network Based Approach

All the above-mentioned works available in open literature are based on HMM technology. All earlier theories of spoken word recognition [53–57] agree to the fact that the spoken word recognition is a complex, multileveled pattern recognition work performed by neural networks of human brain and the related speech perception process can be modeled as a pattern recognition network. Different levels of the language like lexical, semantic, phonemes can be used as the unit of distribution in the model. All the theories proposed that bottom up and top down processes between feature, phoneme, and word level combines to recognize a presented word. In such a situation, ANN models has the greatest potential, where hypothesis can be performed in a parallel and higher computational approach. ANN models are composed of many nonlinear computational elements operating in parallel and arranged in the pattern of biological neural network. The problem of speech recognition inevitably requires handling of temporal variation and ANN architecture like recurrent neural network (RNN), time delay neural network (TDNN) may proven to be handy in such situations. However, ANN-based speech recognition research is still at the preliminary state. A few works based on ANN technology are listed below.

1. Sarkar and Yegnanarayana have used fuzzy rough neural networks for Vowel Classification in a work reported in 1998. This paper [58] has proposed a fuzzy-rough set-based network which exploits fuzzy-rough membership functions while designing radial basis function neural networks for classification.
2. In 2001, Gangashetty and Yegnanarayana have described ANN models for recognition of consonant–vowel (CV) utterances in [59]. In this paper, an approach based on ANN models for recognition of utterances of syllable like units in Indian languages is described. The distribution capturing ability of an autoassociative neural network (AANN) model is exploited to perform nonlinear principal component analysis (PCA) for compressing the size of the feature vector. A constraint satisfaction model is proposed in this paper to incorporate the acoustic-phonetic knowledge and to combine the outputs of subnets to arrive at the overall decision on the class of an input utterance.
3. Khan et al. describe an ANN-based preprocessor for recognition of syllables in 2004. In this work [60], syllables in a language are grouped into equivalent classes based on their consonant and vowel structure. ANN models are used to preclassify the syllables into the equivalent class to which they belong. Recognition of the syllables among the smaller number of cohorts within a class is done by means of hidden Markov models. The preprocessing stage limits the

confusable set to the cohorts within a class and reduces the search space. This hybrid approach helps to improve the recognition rate over that of a plain HMM-based recogniser.

4. In 2005, Gangashetty, Chandra Sekhar, and Yegnanarayana have described an approach for multilingual speech recognition by spotting consonant–vowel (CV) units. The distribution capturing capability of AANN is used for detection of vowel onset points in continuous speech. Support vector machine (SVM) classifier is used as the classifier and broadcast news corpus of three Indian languages Tamil, Telugu, and Marathi is used [61].
5. Paul et al. in 2009 reported a work on Bangla speech recognition using LPC cepstrum features [62]. The self-organizing map (SOM) structure of ANN makes each variable length LPC trajectory of an isolated word into a fixed length LPC trajectory and thereby making the fixed length feature vector to be fed into the recognizer. The structures of the ANN are designed with MLP and tested with 3, 4, 5 hidden layers using the tan sigmoid transfer functions.
6. In 2012, Sunil and Lajish in a work [63] reported a model for vowel phoneme recognition based on average energy information in the zero-crossing intervals and its distribution using multilayer feedforward ANN. They have observed that the distribution patterns of average energy in the zero-crossing intervals are similar for repeated utterances of the same vowels and varies from vowel to vowel and this parameter is used as a feature to classify five Malayalam vowels in the recognition system. From this study, they have shown that the average energy information in the zero-crossing intervals and its distributions can be effectively utilized for vowel phone classification and recognition.
7. Pravin and Jethva recently in 2013 reported a work [64] on Gujrati speech recognition. MFCC of a few selected spoken words is used as feature to train a MLP-based word recognition system.
8. Chitturi et al. reported a work [65] in 2005, where they have proposed an ANN-based approach to model the lexicon of the foreign language with a limited amount training data. The training data for this work consisted of the foreign language with the phone set of three native languages, 49 phones in Telugu, 35 in Tamil, 48 in Marathi, and 40 in US English. The MLP with backpropagation learning algorithm learns how the phones of the foreign language vary with different instances of context. The trained network is capable of deciphering the pronunciation of a foreign word given its native phonetic composition. The performance of the technique has been tested by recognizing Indian accented English.
9. A work [66] by Thasleema and Narayanan in 2012 explores the possibility of multiresolution analysis for consonant classification in noisy environments. They have used wavelet transform (WT) to model and recognize the utterances of consonant–vowel (CV) speech units in noisy environments. A hybrid feature extraction module namely normalized wavelet hybrid feature (NWHF) using the combination of classical wavelet decomposition (CWD) and wavelet packet decomposition (WPD) along with z-score normalization technique is designed

in this work. CV speech unit recognition tasks performed for both noisy and clean speech units using ANN and k Nearest Neighborhood.

10. In 2010, Sukumar et al. has reported a work on recognition of isolated question words of Malayalam language from speech queries using ANN- and discrete wavelet transform (DWT)-based speech feature [67].
11. Sarma et al. has reported works on recognition of numerals of Assamese language in [68, 69] using ANN in 2009. In [68], the ANN models are designed using a combination of SOM and MLP constituting a learning vector quantization (LVQ) block trained in a cooperative environment to handle male and female speech samples of numerals. This work provides a comparative evaluation of several such combinations while subjected to handle speech samples with gender-based differences captured by a microphone in four different conditions viz. noiseless, noise mixed, stressed, and stress-free. In [69], the effectiveness of using an adaptive LMS filter and LPC cepstrum to recognize isolated numerals using ANN-based cooperative architectures. The entire system has two distinct parts for dealing with two classes of input classified into male and female clusters. The first block is formed by a MLP which acts like a class mapper network. It categorizes the inputs into two gender-based clusters.
12. In 2010, Bhattacharjee presented a technique for the recognition of isolated keywords from spoken search queries [70]. A database of 300 spoken search queries from Assamese language has been created. In this work, MFCC has been used as the feature vector and MLP to identify the phoneme boundaries as well as for recognition of the phonemes. Viterbi search technique has been used to identify the keywords from the sequence of phonemes generated by the phoneme recognizer.
13. A work by Dutta and Sarma [71] in 2012 describes a speech recognition model using RNN where linear predictive coding (LPC) and mel frequency cepstral coefficient (MFCC) are used for feature extraction in two separate decision blocks and the decision is taken from the combined architecture. The multiple feature extraction block-based model provides 10 % gain in the recognition rate in comparison to the case when individual feature extractor is used.
14. In 2013, Bhattacharjee in [72] provided a comparative study of linear predictive cepstral coefficients (LPCC) and MFCC features for the recognition of phones of Assamese language. Two popular feature extraction techniques, LPCC and MFCC, have been investigated and their performances have been evaluated for the recognition using a MLP-based baseline phoneme recognizer in quiet environmental condition as well as at different level of noise. It has been reported that the performance of LPCC-based system degrades more rapidly compared to the MFCC-based system under environmental noise condition whereas under speaker variability conditions, LPCC shows relative robustness when compared to MFCC though the performance of both the systems degrade considerably.
15. Sarma and Sarma in 2013, reported a work [73] where a hybrid ANN model is used to recognize initial phones from CVC-type Assamese words. An SOM-based algorithm is developed to segment the initial phonemes from its word

counterpart. Using a combination of three types of ANN structures, namely recurrent neural network (RNN), SOM, and probabilistic neural network (PNN), the proposed algorithm proves its superiority over the discrete wavelet transform (DWT)-based phoneme segmentation.

14.4 Conclusion

It can be concluded from the above literature that the speech recognition technology for Indian languages has not yet covered all the official languages. A few works are done in Hindi language by IBM research lab and a few other research groups have appreciable quality. A few other works have covered Marathi, Tamil, Telugu, Punjabi, Assamese, and Bengali languages which are widely spoken throughout the country. A few works are reported on multilingual speech recognition. However, ASR technologies are yet to be reported in some other mainstream languages like Urdu, Sanskrit, Kashmiri, Sindhi, Konkani, Manipuri, Kannada, Nepali etc. The HMM-based works have already supported the use of continuous speech. In contrast, ANN-based works are still centered around isolated words. But the scenario looks bright and many new success stories shall be reported in near future which shall take ASR technology in Indian languages to new heights.

References

1. Juang BH, Rabiner LR (2004) Automatic speech recognition—a brief history of the technology development. http://www.ece.ucsb.edu/Faculty/Rabiner/ece259/Reprints/354_LALI-ASRHistory-final-10-8.pdf
2. Gales M, Young S (2007) The application of hidden Markov models in speech recognition. Found Trends Sig Process 1(3):195–304
3. Levinson SE, Rabiner LR, Sondhi MM (1983) An introduction to the application of the theory of probabilistic functions of a Markov process to automatic speech recognition. Bell Syst Tech J 62(4):1035–1074
4. Ferguson JD (1980) Hidden Markov analysis: an introduction, in hidden Markov models for speech. Institute for Defense Analyses, Princeton
5. Rabiner LR, Juang BH (2004) Statistical methods for the recognition and understanding of speech. In: Encyclopedia of language and linguistics
6. Davis KH, Biddulph R, Balashek S (1952) Automatic recognition of spoken digits. J Acoust Soc Am 24(6):637–642
7. Suzuki J, Nakata K (1961) Recognition of Japanese vowels preliminary to the recognition of speech. J Radio Res Lab 37(8):193–212
8. Sakai J, Doshita S (1962) The phonetic typewriter, information processing. In: Proceedings IFIP Congress, Munich
9. Nagata K, Kato Y, Chiba S (1963) Spoken digit recognizer for Japanese language. NEC Res Dev (6)
10. Fry DB, Denes P (1959) The design and operation of the mechanical speech recognizer at University College London. J British Inst Radio Eng 19(4):211–229

11. Martin TB, Nelson AL, Zadell HJ (1964) Speech recognition by feature abstraction techniques. Technical Report AL-TDR-64-176, Air Force Avionics Laboratory
12. Vintsyuk TK (1968) Speech discrimination by dynamic programming. Kibernetika 4(2):81–88
13. Viterbi AJ (1967) Error bounds for convolutional codes and an asymptotically optimal decoding algorithm. IEEE Trans Inf Theory 13:260–269
14. Atal BS, Hanauer SL (1971) Speech analysis and synthesis by linear prediction of the speech wave. J Acoust Soc Am 50(2):637–655
15. Itakura F, Saito S (1970) A statistical method for estimation of speech spectral density and formant frequencies. Electr Commun Japan 53(A):36–43
16. Lowerre BT (1976) The HARPY speech recognition system. Doctoral thesis, Carnegie-Mellon University, Department of Computer Science
17. Baker JK (1975) The Dragon system an overview. IEEE Trans Acoust Speech Sig Process 23(1):24–29
18. Jelinek F (1976) Continuous speech recognition by statistical methods. Proc IEEE 64(4):532–556
19. Juang BH (1984) On the hidden Markov model and dynamic time warping for speech recognition a unified view. AT T Tech J 63(7):1213–1243
20. Juang BH (1985) Maximum-likelihood estimation for mixture multivariate stochastic observations of Markov chains. AT T Tech J 64(6):1235–1249
21. Levinson SE, Rabiner LR, Sondhi MM (1983) An introduction to the application of the theory of probabilistic functions of a Markov process to automatic speech recognition. Bell Syst Tech J 62(4):1035–1074
22. Itakura F (1975) Minimum prediction residual principle applied to speech recognition. IEEE Trans Acoust Speech Sig Process 23:57–72
23. Rabiner LR, Levinson SE, Rosenberg AE, Wilpon JG (1979) Speaker independent recognition of isolated words using clustering techniques. IEEE Trans Acoust Speech Sig Process 27:336–349
24. Hu YH, Hwang JN (2002) Handbook of neural network signal processing. In: The electrical engineering and applied signal processing. CRC Press, Boca Raton
25. Lippmann RP (1990) Review of neural networks for speech recognition. In: Readings in speech recognition, pp 374–392. Morgan Kaufmann Publishers, San Mateo
26. Evermann G, Chan HY, Gales MJF, Hain T, Liu X, Mrva D, Wang L, Woodland P (2004) Development of the 2003 CU-HTK conversational telephone speech transcription system. In: Proceedings of ICASSP, Montreal, Canada
27. Matsoukas S, Gauvain JL, Adda A, Colthurst T, Kao CI, Kimball O, Lamel L, Lefevre F, Ma JZ, Makhoul J, Nguyen L, Prasad R, Schwartz R, Schwenk H, Xiang B (2006) Advances in transcription of broadcast news and conversational telephone speech within the combined EARS BBN/LIMSI system. IEEE Trans Audio Speech Lang Process 14(5):1541–1556
28. Languages. The Constitution of India, Eight Schedule, Articles 344(1) and 351:330. Available via http://lawmin.nic.in/coi/coiason29july08.pdf
29. Rabiner L, Juang BH (1986) An introduction to hidden Markov models. IEEE ASSP Mag 3(1):4–16
30. Young S, Kershaw D, Odell J, Ollason D, Valtchev V, Woodland P (2000) The HTK book. http://htk.eng.cam.ac.uk/
31. Lee KF, Hon HW, Reddy R (1990) An overview of the SPHINX speech recognition system. IEEE Trans Acoust Speech Sig Process 38(1):35–45
32. Samudravijaya K, Ahuja R, Bondale N, Jose T, Krishnan S, Poddar P, Rao PVS, Raveendran R (1998) A feature-based hierarchical speech recognition system for Hindi. Sadhana 23(4):313–340
33. Rajput N, Subramaniam LV, Verma A (2000) Adapting phonetic decision trees between languages for continuous speech recognition. In: Proceedings of IEEE international conference on spoken language processing. Beijing, China
34. Kumar M, Rajput N, Verma A (2004) A large-vocabulary continuous speech recognition system for Hindi. IBM J Res Dev 48(5/6):703–715

35. Gaurav DS, Deiv G, Sharma K, Bhattacharya M (2012) Development of application specific continuous speech recognition system in Hindi. J Sig Inf Process 3:394–401
36. Kumar M, Aggarwal RK, Leekha G, Kumar Y (2012) Ensemble feature extraction modules for improved Hindi speech recognition system. Proc Int J Comput Sci Issues 9(3):359–364
37. Bhuvanagirir K, Kopparapu SK (2012) Mixed language speech recognition without explicit identification of language. Am J Sig Process 2(5):92–97
38. Thangarajan R, Natarajan AM, Selvam M (2008) Word and triphone based approaches in continuous speech recognition for Tamil Language. Wseas Trans Sig Process 4(3):76–85
39. Kalyani N, Sunitha KVN (2009) Syllable analysis to build a dictation system in Telugu language. Int J Comput Sci Inf Secur 6(3):171–176
40. Usha Rani N, Girija PN (2012) Error analysis to improve the speech recognition accuracy on Telugu language. Sadhana 37(6):747–761
41. Das B, Mandal V, Mitra P (2011) Bengali speech corpus for continuous automatic speech recognition system. In: Proceedings of international conference on speech database and assessments, pp 51–55
42. Dua M, Aggarwal RK, Kadyan V, Dua S (2012) Punjabi automatic speech recognition using HTK. Int J Comput Sci Issues 9(4):359–364
43. Mehta LR, Mahajan SP, Dabhade AS (2013) Comparative study of MFCC And LPC for Marathi isolated word recognition system. Int J Adv Res Electr Electr Instrum Eng 2(6):2133–2139
44. Udhyakumar N, Swaminathan R, Ramakrishnan SK (2004) Multilingual speech recognition for information retrieval in Indian context. In: Proceedings of the student research workshop at HLT-NAACL, pp 1–6
45. Anumanchipalli G, Chitturi R, Joshi S, Kumar R, Singh SP, Sitaram RNV, Kishore SP (2005) Development of Indian language speech databases for large vocabulary speech recognition systems. In: Proceedings of international conference on speech and computer
46. Lakshmi A, Murthy HA (2008) A new approach to continuous speech recognition in Indian languages. In: Proceedings national conference communication
47. Lakshmi SG, Lakshmi A, Murthy HA, Nagarajan T (2009) Automatic transcription of continuous speech into syllable-like units for Indian languages. Sadhana 34(2):221–233
48. Bhaskar PV, Rao SRM, Gopi A (2012) HTK based Telugu speech recognition. Int J Adv Res Comput Sci Softw Eng 2(12):307–314
49. Mohan A, Rose R, Ghalehjegh SH, Umesh S (2013) Acoustic modelling for speech recognition in Indian languages in an agricultural commodities task domain. Speech Commun. http://dx.doi.org/10.1016/j.specom.2013.07.005
50. Povey D, Burget L, Agarwal M, Akyazi P, Kai F, Ghoshal A, Glembek O, Goel N, Karafiat M, Rastrow A (2011) The subspace Gaussian mixture model A structured model for speech recognition. Comput Speech Lang 25(2):404–439
51. Rose RC, Yin SC, Tang Y (2011) An investigation of subspace modeling for phonetic and speaker variability in automatic speech recognition. In: Proceedings of IEEE international conference on acoustics, speech, and signal processing
52. Mohan A, Ghalehjegh SH, Rose RC (2012) Dealing with acoustic mismatch for training multilingual subspace Gaussian mixture models for speech recognition. In: Proceedings of IEEE international conference on acoustics, speech and signal processing
53. Diehl RL, Lotto AJ, Holt LL (2004) Speech perception. Annu Rev Psychol 55:149–179
54. Eysenck MW (2004) Psychology-an international persepective. Psychology Press. http://books.google.co.in/books/about/Psychology.html?id=l8j_z5-qZfACredir_esc=y
55. Jusczyk PW, Luce PA (2002) Speech perception and spoken word recognition: past and present. Ear Hear 23(1):2–40
56. Bergen B (2006) Linguistics 431/631: connectionist language modeling. Meeting 10: speech perception. http://www2.hawaii.edu/bergen/ling631/lecs/lec10.htm
57. McClelland JL, Mirman D, Holt LL (2006) Are there interactive processes in speech perception? Trends Cogn Sci 10(8):363–369
58. Sarkar M, Yegnanarayana B (1998) Fuzzy-rough neural networks for vowel classification. IEEE Int Conf Syst Man Cybern 5:4160–4165

59. Gangashetty SV, Yegnanarayana B (2001) Neural network models for recognition of consonant-vowel (CV) utterances. In: Proceedings of international joint conference on neural networks
60. Khan AN, Gangashetty SV, Yegnanarayana B (2004) Neural network preprocessor for recognition of syllables. In: Proceedings of international conference on intelligent sensing and information processing
61. Gangashetty SV, Sekhar CC, Yegnanarayana B (2005) Spotting multilingual consonant-vowel units of speech using neural network models. In: Proceeding of international conference on non-linear speech processing
62. Paul AK, Das D, Kamal M (2009) Bangla speech recognition system using LPC and ANN. In: Proceedings of the seventh international conference on advances in pattern recognition, pp 171–174
63. Sunil KRK, Lajish VL (2012) Vowel phoneme recognition based on average energy information in the zerocrossing intervals and its distribution using ANN. Int J Inf Sci Tech 2(6)
64. Pravin P, Jethva H (2013) Neural network based Gujarati language speech recognition. Int J Comput Sci Manage Res 2(5)
65. Chitturi R, Keri V, Anumanchipalli G, Joshi S (2005) Lexical modeling for non native speech recognition using neural networks. In: Proceedings of international conference of natural language processing
66. Thasleema TM, Narayanan NK (2012) Multi resolution analysis for consonant classification in noisy environments. Int J Image Graph Sig Process 8:15–23
67. Sukumar AR, Shah AF, Anto PB (2010) Isolated question words recognition from speech queries by using artificial neural networks. In: Proceedings of international conference on computing communication and networking technologies, pp 1–4
68. Sarma MP, Sarma KK (2009) Assamese numeral speech recognition using multiple features and cooperative LVQ-architectures. Int J Electr Electr Eng 5(1):27–37
69. Sarma M, Dutta K, Sarma KK (2009) Assamese numeral corpus for speech recognition using cooperative ANN architecture. World Acad Sci Eng Technol 28:581–590
70. Bhattacharjee U (2010) Search key identification in a spoken query using isolated keyword recognition. Int J Comput Appl 5(8):14–21
71. Dutta K, Sarma KK (2012) Multiple feature extraction for RNN-based assamese speech recognition for speech to text conversion application. In: Proceedings of international conference on communications, devices and intelligent systems (CODIS), pp 600–603
72. Bhattacharjee U (2013) A comparative study Of LPCC and MFCC features for the recognition of assamese phonemes. Int J Eng Res Technol 2(1):1–6
73. Sarma M, Sarma KK (2013) An ANN based approach to recognize initial phonemes of spoken words of assamese language. Appl Soft Comput 13(5):2281–2291

Chapter 15
Recent Trends in Power-Conscious VLSI Design—A Review

Manash Pratim Sarma

Abstract Power-aware or power-conscious very-large-scale integration (VLSI) design is gradually evolving to be one of the most pronounced and important, yet challenging aspect of system design and implementation. The issue of power is being addressed by different researchers around the globe at different levels of implementation and fabrication. This chapter highlights the recent works in circulation in open literature primarily reported during the present decade. It focuses broadly on four important aspects viz. device design, circuit design, high data rate applications, and synthesis and scheduling. The objective is to discuss the relevant developments as well as to identify existing challenges faced by the community of researchers engaged in power-aware VLSI design.

Keywords Very-large-scale integration (VLSI) · High data rate applications · Power-aware VLSI design · CMOS circuit

15.1 Introduction

Since the invention of transistor, the world has witnessed tremendous changes in electronics design with emphasis on increase in miniaturization and decrease in power requirements. Since the formative years of electronics design, the pace and the trends have established it to be the fastest growing technology ever. With the beginning of solid-state era, the technology roadmap of very-large-scale integration (VLSI) design has always been to increase the packing densities, increasing processing speed and lifetimes, optimization of packaging, and lowering of attrition rates and unit power consumption. With increasing stress on miniaturization and packing densities, power issues have become more crucial. This chapter intends to point out the relevant developments as well as to identify existing challenges and discuss some of the significant contributions made by research groups in power-aware VLSI design.

M.P. Sarma (✉)
Department of Electronics and Communication Engineering, Gauhati University, Guwahati, Assam, India
e-mail: manashpelsc@gmail.com

K.K. Sarma et al. (eds.), *Recent Trends in Intelligent and Emerging Systems*, Signals and Communication Technology, DOI 10.1007/978-81-322-2407-5_15

The rest of the paper is organized as follows: In Sect. 15.2, we discuss about the current trends in power conscious designs with stress on design level modifications. In Sect. 15.3, design level techniques are highlighted. Current trends in high data rate designs are covered in Sect. 15.4. Section 15.5 includes the trends in synthesis aspect of VLSI design. The paper is concluded in Sect. 15.6.

15.2 Device-Level Modifications

The power consideration of VLSI systems starts at the device level. The miniaturization of devices to submicron or nanoscale is beneficial not only in terms of compactness but also to have low-power designs. There are different bottlenecks to achieve low-power condition which are independent of the circuit-level or system-level designs. The minimum effective channel length possible to avoid tunneling in semiconductor designs, primarily with regards to unipolar devices, to happen is a very critical issue which has been partially addressed by having gradual junction near the channel and using multiple gates. Also there happens to be different problems like threshold voltage shift, bulk punch through, gate-induced barrier lowering (GIDL), impact ionization etc. in case of smaller devices [1]. Again the minimum switching energy that a device can transfer is directly related to the effective channel length, hence optimization is required. Different works have been carried out suggesting device-level modifications or employing different techniques to a device to achieve proper performance with considerably lower power. A few such works are discussed here.

1. A design methodology of micro-power analog complementary metal oxide semiconductor (CMOS) cell is presented in [2] where an investigation of the drawbacks of biasing the CMOS in weak inversion region is carried out. It will be worth noting that biasing a CMOS cell in the weak inversion region makes the power consumption to reduce than biasing in the strong inversion region. Weak Inversion region signifies the operating region below the subthreshold voltage. The drain current in weak inversion region is dominated by the diffusion current which yields due to the thermally generated minority carriers under the influence of the gate electric field. Hence, there is a strong dependence of bulk temperature on the drain.
   Analyzing the device behavior and mathematical formulations, we can say that in weak inversion the minimum voltage required to have saturation is less than in strong inversion. But as the equation shows that the W/L ratio will increase which indicates a trade-off between power and area. Since the transconductance is linearly related to drain current, a MOS transistor offers maximum transconductance in the weak inversion and gradually decreases as it moves toward strong inversion. This work also validates weak inversion-based circuits with the use of EKV model (Enz-Krummenacher-Vitoz) which is claimed to be more precise. Basic CMOS Op-Amp and operational transconductance amplifier (OTA) is

modeled and the performance of OTA is seen to better. But the circuit performance is dependent on the stability of the current source used for biasing. Finally, the proposed technique can be viewed to be a successful technique but the improvement of the BW is not addressed.

2. In the paper [3], different techniques of low-voltage, low-power CMOS design in the deep submicron ICs have been discussed. In CMOS devices, though reducing the supply voltage leads to power reduction, the performance gets suffered. Hence, scaling down the threshold level could be a solution but it too influences the dramatic increase in the leakage current which is a great concern in high-performance circuits. The CMOS circuit is characterized by two types of power dissipation: dynamic and static power dissipation during the active mode of operation. In the standby mode, the power dissipation is solely due to the standby leakage current. Again dynamic power dissipation has two components: switching power due to charging and discharging of load capacitance and the short circuit power due to the nonzero rise and fall time of input waveforms. Investigating and analyzing different phenomena responsible for leakage current like reverse bias junction leakage, subthreshold leakage, GIDL, DIBL, Gate oxide tunneling, it was concluded that silicon on insulator (SOI) provides the best result in combating leakage problem. A double gate (DG) fully depleted (FD) CMOS is proved to have excellent immunity to short channel effect and high drive current which makes it an attractive candidate for low-voltage, low-power, and high-performance CMOS design.
3. Multi-gate-length (MGL) and dual-gate-length (DGL) biasing techniques are investigated in [4] for timing constraint-aware active mode leakage power reduction of VLSI circuits. For active mode power reduction, gate lengths above sun-nominal length was proposed since it leads to reduction in run-time leakage. Also gate length biasing can be implemented either in cell level or in transistor level. The DGL approach provides nominal gate-length and one bias value, whether in MGL we may have different bias value. While choosing the gate lengths, the footprint equivalence and interchangeability between cell masters and cell variants must be assured. The authors benchmark cell-level MGL over DGL with the comparison and analysis of leakage power reduction and full chip leakage power variation. A sensitivity-based cell delay and leakage modeling is adopted and finally a Monte Carlo-based leakage reduction investigation was performed. Conclusion was that MGL provides modest to nominal leakage power reduction.

15.3 Circuit-Level or Design-Level Techniques

The circuit-level or design-level techniques to reduce power plays a crucial role in any VLSI circuit implementation. Different optimization issues that lead to the trade offs between power and inherent circuit properties needs to be analyzed properly. Decreasing the bias voltage decreases the overall power but the limit of the minimum bias voltage to be able to properly distinguish between the two logic levels should be known because it is inherent to the circuit [1]. Again the minimum switching energy per cycle and intrinsic gate delay are also circuit specific and also dependent on the bias voltage and capacitance. The global interconnects being an RC network makes a decisive role since the circuit response time is influenced by it. Resolving these sort of numerous issues with a goal to have a low-power, high-performance circuit have been the thrust area of the researchers in the last decade. Such issues are discussed below.

1. An area-conscious, low-voltage controlled, 8T Static Random Access Memory (SRAM) cell has been presented in [5]. The design was also verified in a 0.42 v operated dynamic voltage scaling (DVS) environment in 90 nm technology. The final product comes out to be a 64 kb, 8T-SRAM chip in 90 nm technology with write-back scheme and improved Bit Error Rate (BER) performance.
2. Advanced VLSI architectures like multicore processors, system on chip (SOC), network on chip (NOC) etc. face the problem of high power dissipation and sharp thermal gradient. DVS environment offers power management techniques in such scenario. But the need of converters in such hardware leads to a very bulkier design. An answer to that problem is a single-inductor multiple-output (SIMO) converter. A delay-locked loop (DLL)-based SIMO converter with locking range of 350 ns and simulated in IBM 130 nm has been reported in [6]. A low-power, low-voltage topology is chosen for power management with time multiplexing control. A novel control technique is used which involves utilizing the inherent phase locking property of the DLL to lock onto the converter's regulation error between the output voltages and their corresponding references. Maximum efficiency observed was 87.2 %.
3. Since process variation is a major concern in CMOS circuit manufacturing, scaling down the critical dimensions leads to front-end of line (FEOL) systematic variations to be a major performance variability contributor. The work [7] proposes a leakage aware methodology to improve the circuit performance in terms of standby static power and power delay product (PDP).
4. With the growing need of portable handheld electronic equipments, ultra-low-power (ULP) circuit designs are at the prime demand. Subthreshold operation of devices has proved to be an efficient candidate for ULP circuits, but with increase in interconnect density, delay and cross-talk also increase significantly and is more prominent in nanoscale circuits. In the paper [8], the suitability of interconnect strategies has been studied and an new interconnect optimization technique is devised to have subthreshold operations at reduced cross-talk without compromising the performance. They have considered the longest interconnect length for driver optimization of global interconnect and also the total path delay under subthreshold is determined by the driver resistance and the interconnect capacitance.
5. The use of dynamic logic is widespread due to its numerous advantages like less transistor count and high speed which results for reduced load capacitance. A pseudodynamic buffer (PDB) in the footed dynamic logic has been proposed in [9] for both pull-up and pull-down network to reduce static power at 0.35 m CMOS technology. The propagation of precharge pulse to the output stage in

dynamic logic increases the static power and hence the propagation was stopped to get a low power performance. Moreover, the proposed structure eliminates the necessity of having a clock-controlled transistor resulting a more compact design.

6. A 65-nm CMOS technology-based 7T SRAM architecture with high data scalability and yield is presented in [10]. Ground gating by selective control of column virtual ground significantly reduces the leakage power consumption. Also the proposed architecture have 30 % higher soft error critical charge in comparison to its 6 T counterpart. Also the proposed sense amplifier can be effectively used with 7 T structure with similar dimension as the bit-cell.
7. A novel mechanism of reducing overall power consumption in nanoscale circuits with the modification in the reference circuit is suggested in [11]. The simulation is done at 65 nm CMOS technology with the use of transistor stacking and finally reduces the power consumption at the expense of 10 % increase of delay.

15.4 Trends in High Data Rate Applications

With ever increasing need of highly integrated miniaturized portable equipments with more functionality, high speed, and low power have become two decisive ingredients. High-performance processors most often are required to be embedded with deep pipelining [12], which substantially increases the power consumption. Likewise, power densities also become crucial with scaling down of the technologies. So optimality is not only associated with speed but also with power and hence consideration of both speed and power is important to cope up with advanced high data rate strategies. Apart from algorithm-level strategies there may be the requirement of circuit-level techniques like use of dynamic circuits only when speed demand is prime, use of delayed-reset domino circuit, pulse to static conversion at the output without sacrificing speed etc. Extensive research is being carried out and few important works are discussed here.

1. An ultra-low power multimedia processor with Motion Pictures Experts Group (MPEG) 4 encoding for mobile applications is presented in [13]. Optimization was suggested from algorithmic level to hardware level with an highly accurate, reduced computational complexity algorithm.
2. In 2003, it was predicted [14] that the dynamic branch prediction problem can be addressed well with the use of Artificial Neural Networks (ANN). They were verifying the learning vector quantization (LVQ) and back propagation network with the achievement of lower misprediction and were suggesting that ANNs would be interesting candidates for these sort of problems. But the power related issue was not discussed.
3. A power-aware ANN-based branch prediction scheme to incorporate anti-aliasing technique has been discussed in [15]. The scheme dynamically learns to dissipate less power during successive calls making a distinction with previous techniques of hardware-level, power-aware designs. The ANN BP operation is designed

as a combination of parallelism and pipelining to avoid degradation of timing. Misprediction is reduced from 11.1 to 8.5 % and is claimed to be the first power-aware BP.

4. Finite field inversion is computationally most intensive operation. Frequently used algorithms to have inverse binary field include Extended Euclidean Algorithm (EEA), Binary Euclidean Algorithm (BEA), Montgomery Inversion Algorithm (MIA), and Itoh-Tsujii algorithm (ITA). In terms of hardware implementation, ITA is said to be efficient for its higher speed due to the presence of the field multiplier. In [16], a generalization of parallel ITA structure is adopted but is seen to be suffering from a limited speed. Hence, a new sequential architecture is suggested to reduce critical delay without increasing the clock cycle requirement. It is either a cascade of 2-n circuits or a cascade of 2-n circuits and is tested in different FPGA devices. The final outcome is an increase in speed by 15 % for inverse computation compared to a quad ITA implementation and 80 % compared to the square and square-root operation-based parallel ITA.
5. The Elliptic Curve Cryptosystem (ECC)-based processor is becoming an attractive candidate in recent years and extensive research work is going on in this area. The Galois Fields (GF(2m)) find its importance in hardware implementation due to carry free arithmetic. One such GF(2m) compatible, flexible, area efficient, low-power, high-speed, versatile bit-serial multiplier design is reported in [17]. It works in the most significant bit (MSB)-first mechanism for different operand lengths. The evaluation of efficiency is based on latency, critical path, and space complexity.

15.5 Synthesis and Scheduling

Synthesis and scheduling is a crucial issue related to testing and testability of the chips. Successful testing necessitates the power issues to be addressed properly. We are presenting here some of such works related to power-aware synthesis.

1. Built-in self test (BIST) increases circuit activities and thereby giving rise to power in data path circuits leading to yield and reliability problem. The voltage drop during BIST makes many good circuits to fail, therefore testing of low-power chip poses critical issues which need to be addressed. A power-conscious test synthesis and scheduling is introduced in [18] with targeted dataflow-intensive applications.
2. In all highly scaled VLSI circuits, high temperature has a serious impact on circuit performance which results due to the decrease in switching speed as the interconnect resistance increases. A thermal management strategy in the architectural synthesis flow at multiple stages is proposed in [19]. This temperature-aware high-level synthesis reduces peak temperature by 7.34° C compared to temperature-unaware designs and thereby saving leakage power.

3. A low-power thermally aware algorithm for behavioral synthesis with the reduction in temperature by 23 % and chip area by 10 % is presented in [20]. To make efficient sampling of multi-objective solution space possible and to have an inexpensive thermal analysis, two-stage annealing is adopted. Finally, there is an decrease in peak module temperature by 15 % with less than 1 % average power difference.
4. In the present scenario of power-hungry multicore processors, design of power-aware NOCs is challenging. A framework of automated power level synthesis of regular NOCs is presented in [21]. They have reduced the network traffic to 62 % with the use of heuristics. The partitioning technique optimizes both communication power as well as computation power while using distributed decision-making mapping scheme over the whole mesh. Also routing path allocation algorithm integrates link-insertion and routing. There was a saving of 32 % of communication power and 13 % of overall power.

15.6 Conclusion

In this chapter, we have seen that there is a continuous work to achieve the objective of designing high-performance VLSI systems at lowest possible power levels. The trend observed in recent years is more inclined toward power-aware designs. Researchers have found ways to lower the power without violating the inherent limits and making breakthrough enabling design of high-performance portable computing and communication equipments for multiple applications.

References

1. Roy K, Prasad SC (2011) Low-power CMOS VLSI circuit design. Wiley India (P) Ltd, India
2. Pimental J, Salazar F, Pacheco M, Vellasco M, Gavriel Y (2000) Very low-power analog cells in CMOS. In: Proceeding of 43rd IEEE midwest symposium on circuits and systems, Lansing MI, pp 328–331
3. Liqiong W, Roy K, De VK (2000) Low voltage low power CMOS design techniques for deep submicron ICs. In: Proceeding of 13th international IEEE conference on VLSI Design, pp 24–29
4. Chen Q, Tirumala S (2009) Multi-gate-length versus dual-gate-length biasing for active mode leakage power reduction: benchmarking and modeling. In: Proceeding of IEEE 8th international conference on ASIC, pp 781–784
5. Morita Y, Fujiwaral H, Noguchil H, Iguchil Y, Niil K, Kawaguchi H, Yoshimotol M (2007) An area-conscious low-voltage-oriented 8T-SRAM design under DVS environment. In: Proceeding of IEEE Symposium on VLSI Circuits, pp 256–257
6. Bondade R, Ma D (2009) A DLL-regulated SIMO power converter for DVS-enabled power-aware VLSI systems. In: Proceeding of 52nd IEEE international midwest symposium on circuits and systems, pp 961–964

7. Aghababa H, Afzali-Kusha A, Forouzandeh B (2009) Static power optimization of a full-adder under front-end of line systematic variations. In: Proceeding of 4th IEEE international design and test workshop, pp 1–6
8. Pable SD, Hasan H (2012) Ultra-low-power signaling challenges for sub-threshold global interconnects. Integr VLSI J ELSEVIER 45(2):186–196
9. Tang F, Bermak A, Gu Z (2012) Low power dynamic logic circuit design using a pseudo dynami cbuffer. Integr VLSI J ELSEVIER 45(4):395–404
10. Hussain W, Jahinuzzaman S (2012) A read-decoupled gated-ground SRAM architecture for low-power embedded memories. Integr VLSI J ELSEVIER 45(3):229–236
11. Chakraborty A, Pradhan SN (2012) A technique for power reduction of CMOS circuit at 65nm technology. In: Proceeding of 1st international conference on recent advances in information technology, RAIT, pp 576–580
12. Takahashi O, Dhong SO, Hofstee P, Silberman J (2001) High-speed, power-conscious circuit design techniques for high-performance computing. In: Proceeding of technical papers and international symposium on VLSI technology, systems, and applications, pp 279–282
13. Alfonso D, Artieri A, Capra A, Mancuso M, Pappalardo F, Rovati F, Zafalon R (2002) Ultra low power multimedia processor for mobile multimedia applications. In: Proceeding of 28th European solid-state circuits conference, ESSCIRC, pp 63–69
14. Egan C, Steven G, Quick P, Anguera R, Steven F, Vintan L (2003) Two level branch prediction using neural networks. J Syst Archit ELSEVIER 49(12–15):557–570
15. Sethuram R, Khan OI, Venkatanarayanan VH, Bushnell ML (2007) A neural net branch predictor to reduce power. In: Proceeding of 20th IEEE international conference on VLSI design, pp 679–684
16. Roy SS, Rebeiro C, Mukhopadhyay D (2012) Generalized high speed Itoh-Tsujii multiplicative inversion architecture for FPGAs. Integr VLSI J ELSEVIER 45(3):307–315
17. Zakerolhosseini A, Nikooghadam M (2013) Low-power and high-speed design of a versatile bit-serial multiplier in finite fields GF(2m). Integr VLSI J ELSEVIER 46(2):211–217
18. Nicolici N, Al-Hashimi BM (2003) Power-conscious test synthesis and scheduling. In: Proceeding of IEEE international test conference (ITC), pp 48–55
19. Mukherjee R, Memik SO (2006) An integrated approach to thermal management in high-level synthesis. IEEE Trans Very Large Scale Integr (VLSI) Syst 14(11):1165–1174
20. Krishnan V, Katkoori S (2009) Simultaneous peak temperature and average power minimization during behavioral synthesis. In: Proceeding of 22nd IEEE international conference on VLSI design, pp 419–424
21. Kapadia N, Pasricha S (2012) A framework for low power synthesis of interconnection networks-on-chip with multiple voltage islands. Integr VLSI J ELSEVIER 45(3):271–281

Chapter 16
Application of Soft Computing Tools in Wireless Communication—A Review

Kandarpa Kumar Sarma

Abstract The proliferation of number of users in a limited wireless spectrum have raised the levels of inter symbol interference (ISI) and have also contributed towards probable degradation of quality of service (QoS). The key challenges faced by upcoming wireless communication systems is to provide high-data-rate wireless access with better QoS. Also, the fast shrinking spectrum for such communication have necessitated the development of methods to increase spectral efficiency. Multiple input multiple output (MIMO) wireless technology is a viable option in such a situation and is likely to be able to meet the demands of these ever-expanding mobile networks. Many researchers have explored this field over a considerable period of time. A sizable portion of the research have been on the application of traditional statistical methods in such areas. Over the years, soft computational tools like artificial neural network (ANN), fuzzy systems and their combinations have received attention in the diverse segments of wireless communication. This is because of the fact that these are learning based systems. These learn from the environment, retain the knowledge and use it subsequently. This paper highlights some of the important application areas in wireless communication which have reported the use of soft computing tools in wireless communication that are in circulation in open literature.

Keywords Multiple input multiple output (MIMO) technology · Soft computation · Wireless communication · Multi layer perceptron (MLP) · Recurrent neural network (RNN) · Fuzzy · Fuzzy-neural · Neuro-fuzzy

16.1 Introduction

The proliferation of mobile communication networks over the last decade has increased the use of the wireless spectrum in exponential terms. Increase in number of users in a limited spectrum have raised the levels of inter symbol interference

---

K.K. Sarma (✉)
Department of Electronics and Communication Technology,
Gauhati University, Guwahati 781014, Assam, India
e-mail: kandarpaks@gmail.com

K.K. Sarma et al. (eds.), *Recent Trends in Intelligent and Emerging Systems*,
Signals and Communication Technology, DOI 10.1007/978-81-322-2407-5_16

(ISI) and also have increased the possibility of degraded quality of service (QoS). The key challenge faced by upcoming wireless communication systems is to provide high-data-rate wireless access at better QoS. Also, destructive addition of multipath components within the fast shrinking spectrum available for wireless communication have necessitated the development of methods to increase spectral efficiency and explore innovative solutions. Additionally, there is a constant demand for higher bandwidth, increased data rates, lower cost, greater coverage etc. for which the mobile networks are creating congestion in the available spectrum. Multiple-input multiple-output (MIMO) wireless technology is a viable option in such a situation and is likely to be able to meet the demands of these ever-expanding mobile networks. Many researchers have explored this field over a considerable period of time. A sizeable portion of the research have been on the application of traditional statistical methods in such areas. Over the years, soft computing tools like artificial neural network (ANN), fuzzy systems and their combinations have received attention in wireless communication. This is because of the fact that these are learning based systems. These learn from the environment, hold back the learning and use it subsequently. This paper highlights some of the important application areas in wireless communication which have reported the use of soft computing tools in wireless communication. As MIMO is a viable option to meet the demands of expanding mobile networks, a larger section of the research have focused on the application of soft computing tools in this area as well.

The rest of the paper is organized as follows. In Sect. 16.2, the importance of soft-computing tools in wireless communication is highlighted. We discuss about the application of the ANN in feedforward form in Sect. 16.3. Application of recurrent structures in wireless communication is discussed in Sect. 16.4. The important works reported in the domain of fuzzy based applications in wireless communication are included in Sect. 16.5. The work is concluded in Sect. 16.6.

16.2 Importance of Soft Computing Tools in Wireless Communication

Soft computing tools like ANN, fuzzy systems and their combinations have become important segments of systems related to wireless communication. This is because of the fact that these being learning based systems, are better placed to use channel side information (CSI) for improved performance. ANNs have already received considerable attention as an optional technique for equalization and other such applications in wireless communication. The most preferable aspects of the ANN in these applications have been parallelism, adaptive processing, self-organization, universal approximation and ability of tackling highly nonlinear problems. Also, as the ANN learn complex patterns, it acts as a reliable estimator and hence is used for the modeling a host of phenomena observed in wireless systems and MIMO channels.

ANN uses model free data to capture changes [1] which have been effectively used in design of systems for channel estimation and symbol recovery simultaneously.

On the other hand, fuzzy systems are rule driven tools suitable for uncertain conditions executing minute regulations for improved control and decision-making. This is somewhat different to that observed in ANNs. ANNs are nonparametric prediction tools that have the ability to replicate biological cognitive behaviour but cannot explain to the user how the system derives a decision. Hidden knowledge in ANN is not associated with a single aspect of a given problem. Fuzzy systems, on the other hand, attempt to extract expert-level knowledge embedded in a process. The rule generation process is critical in extracting knowledge and arrive at decisions from near random or unknown situations. Fuzzy systems are applicable where sufficient expert knowledge about a process in available while ANN is comfortable with situations that have sufficient process data. Therefore, while ANNs have numeric-quantitative capability, fuzzy systems exhibit symbolic-qualitative capacity. Thus, hybrid systems formed by combinations of ANN and fuzzy methods have adaptability, parallelism, non-linear processing, robustness and learning in data rich environment acquired from ANNs and modeling uncertainty and qualitative knowledge related to fuzzy systems. It provides neuro-fuzzy (NF) or fuzzy-neural (FN) systems the ability to acquire numeric-qualitative expert-level decision-making and demonstrate greater adaptability and robustness while handling unknown process or situations.

16.3 Application of Feedforward ANN and MLP in Wireless Communication

ANNs are available in a host of forms [1]. In the rudimentary feedforward form, the ANN is configured as a multi layer perceptron (MLP) which is trained by (error) back propagation (BP) algorithm [1]. In this section, we highlight some of the important applications of ANN in feedforward form related to wireless communication.

For single input single output (SISO) and single input multi output (SIMO) setups, MLPs have been extensively used. As an extension to the application of SISO and SIMO, MIMO systems also have attracted considerable employment of ANN for a range of situations like channel equalization, interference cancellation, identification and estimation. A few such works are included below:

1. A three layer ANN along with feedback is used for MIMO channel estimation and equalization and is reported in [2]. The work uses a Kalman filter and a feedforward ANN to perform MIMO channel estimation.
2. Another work cited in [3] reports the application of ANN for location estimation and CCI suppression in cellular networks.
3. A work related to blind equalization of a noisy channel by linear ANN is reported in [4].
4. Another work of similar nature is available as cited in [5] where blind channel equalization and estimation is performed using ANN. This work discusses

application of ANN mainly with a time invariant SISO channel. Along with CCI cancelation and equalization, estimation of MIMO channels have also received attention with regards to application of ANN and related tools.

5. A method based on iterative estimation of MIMO channels using support vector machine (SVM) is reported in [6].
6. Another work [7] reports the application of singular value decomposition (SVD) based adaptive channel estimation for MIMO-OFDM systems.
7. SVD has also been used for subchannel CCI cancelation in a MIMO system as described in the work cited in [8].
8. Like SVD, ANN has always been a preferred tool for developing applications in high data rate systems like MIMO-OFDM systems. Following these applications, identification of nonlinear MIMO channels using ANNs has also been reported. One such work is [9].
9. The work [10] reports the use of a MLP for multipath Rayleigh channel estimation in a MIMO set-up.
10. Works of similar nature that deals with static and slowly varying MIMO channels has already been reported. A work [11], reports the use of ANN for MIMO-OFDM channel estimation. The work uses pilots to estimate the channel impulse response based on LS criteria. To improve the estimation performance, an ANN approach is applied to track the variations of the channel using a variable step-size RLS.
11. Application of Multi-ADAptive LINear Element (MADALINE) which is a feed-forward ANN, for parameter estimation of linear time invariant (LTI) MIMO systems is reported in [12].
12. As channel estimation is a time varying phenomenon, dynamic ANNs are more suitable. The work [13] reports the use of a dynamic ANN topology for MIMO-OFDM systems. The MLPs are tested to check their ability to perform symbol recovery as well under fluctuating conditions shown by the MIMO channels.
13. MLPs capturing time—varying patterns of input data must have temporal characteristics [14] which can be developed by building memory into an ANN [1]. There are two basic methods which can be used to introduce memory into an ANN. The first one is to introduce time delays in the ANN and to adjust parameters during learning phase. The second way is to use positive feedback which can make the ANN recurrent. Recurrent networks use global feed-forward and local feedback sections [1].

In the above cases we have seen a number of reported works which have used ANN in feedforward form to deal with certain phenomenon observed in wireless channels. In these cases it is seen that an ANN can be specially configured to mitigate some of the deficiencies of multi-user transmission. The advantage of these schemes is that no pilot symbol bits are required to be inserted with such transmissions which can contribute towards preserving bandwidth and increasing spectral efficiency.

16.4 Application of Recurrent Neural Network (RNN) in Wireless Communication

Another form of ANN is recurrent neural network (RNN) which possesses better ability to deal with time varying systems than the MLP. This is because of the fact that the RNN has local and global feedback paths which enable contextual processing of applied samples [1]. RNNs have also been used for a diverse range of applications in wireless communication. Some of the relevant literature are cited between [15–29].

1. Blind equalization has been the most common area in which RNNs have been applied. A work cited in [15] uses a RNN for blind equalization which proves to be effective.
2. A more extensive application of the RNN is observed in case of a system developed for MIMO channel prediction using a particle swarm optimization (PSO)-evolutionary algorithm (EA)-differential evolution PSO (DEPSO) (PSO-EA-DEPSO) off-line training algorithm. This predictor is shown to be robust to varying channel scenarios [16]. The work only concentrates on MIMO channels and is not directed towards recovery of data symbols transmitted through the channel.
3. Another related work cited in [17] improves the effort reported in [16] by using an on-line approach. A new hybrid PSO-EA-DEPSO algorithm is presented for training a RNN for MIMO channel prediction. This algorithm is shown to outperform RNN predictors trained off-line by PSO, EA, and DEPSO as well as a linear predictor. This work also is not directed towards recovery of transmitted signal content. Further, the work doesn't specify if real and complex signals are considered in split form or in coupled form.
4. A work [18] provides a new adaptive neural predictor for GPS jamming suppression applications designed using the efficient square-root extended Kalman filter (SREKF) algorithm to adjust the synaptic weights in a RNN architecture and thereby estimate the stationary and non-stationary narrowband/FM waveforms.
5. A novel approach to adaptive channel equalization with RNN for a QSPK signal constellation is given in [19]. The work deals with wireless communications in non linear channels for M-PSK and M-QAM modulation schemes.
6. RNNs with extended Kalman filter (EKF) algorithm has also been used for nonlinear equalization in satellite communication. This is reported by a work as indicated by [20].
7. Another work cited in [21] applied complex real time recurrent learning fully RNN extended Kalman filter trained (CRTRLEKF) in adaptive equalization for cellular communications. Results illustrate the strength of the method in wide sense stationary-uncorrelated scattering (WSS-US) channel model.
8. Accurate and timely estimation of CSI will guarantee the QoS by admission control, inter and intra network handovers in non line of sight (NLOS) channels. For such a problem, bit error rate (BER) is predicted by two different RNN architectures such as recurrent radial basis function network (RRBFN) and echo state network (ESN) [22].

9. RNNs trained with gradient-based algorithms such as real-time recurrent learning (RTRL) or back-propagation through time (BPTT) have a drawback of slow convergence rate. A derivative-free Kalman filter, also called the unscented Kalman filter (UKF), for training a fully connected RNN is presented for nonlinear equalization in [23, 24].
10. A work on the use of Kalman filter-trained RNN equalizers for time-varying channels is reported in [25].
11. A decision feedback RNN based equalization with fast convergence rate for time-varying channels is described in [26].
12. In the work described in [27], the application of fully connected RNNs (FCRNNs) is investigated in the context of narrow-band channel prediction using three different algorithms, namely the RTRL, the global extended Kalman filter (GEKF) and the decoupled extended Kalman filter (DEKF). The system is designed for training the RNN—based channel predictor. The work shows that GEKF and DEKF training are faster than the RTRL based learning.
13. A new method for pruning the complex bilinear recurrent neural network (CBLRNN) using genetic algorithm is proposed in [28] and is applied to equalization of wireless asynchronous transfer mode (ATM) channels.
14. Use of Kalman filter and RNN in hybrid form is reported in [29].

The most preferable aspects of the RNN in these applications have been parallelism, adaptive processing, self-organization, universal approximation and ability of tracking highly nonlinear problems. Here too, the reported works of application of RNN for wireless and MIMO channel modeling have not crossed the traditional limits of experimenting with the training–testing realm. In particular, no reported works have dealt with architectural expansion of the RNN which would have been a natural modification over traditional RNN structures. The training time complexity and design issues are important challenges observed in these works.

16.5 Application of Fuzzy, Fuzzy-Neural and Neuro-Fuzzy Systems in Wireless Communication

Fuzzy systems provide expert-level knowledge for control and decision-making while ANNs are non-parametric prediction tools that have the ability to replicate biological behaviour. Therefore, while ANNs have numeric-quantitative capability, fuzzy systems exhibit symbolic-qualitative capacity. It provides fuzzy-based systems the ability to acquire numeric-qualitative expert-level decision-making and demonstrate greater adaptability and robustness while handling unknown process or situations. These attributes of fuzzy based systems in combination with ANNs have been explored and configured for wireless channel modeling and related phenomenon. Fuzzy and related hybrid systems namely FN and NF systems provide adaptive expert-level decision-making capacity, hence are suitable for a wide range of applications. Fuzzy based systems are efficient tools to be utilized in problems for which either information or knowledge of all factors is insufficient or impossible to obtain. Fuzzy and related hybrid systems have also received attention for application in wireless communication.

1. A work cited in [30] is a must read description for all fuzzy related implementation. The description provides an exhaustive survey of neuro-fuzzy rule generation algorithms together under a unified soft computing framework. The work includes some important works related to rule generation of NFS and relates them to certain real world applications. Another work of similar nature is [31].
2. Some nobel efforts have been put into a publication cited in [32], which contains a survey of fuzzy logic applications and principles in wireless communications. It is reported with the aim of highlighting successful usage of fuzzy logic techniques in telecommunications and signal processing. The authors claimed this is to be the first such study of its kind. This paper focuses firstly on discerning prevalent fuzzy logic or fuzzy-hybrid approaches in the areas of channel estimation, equalization and decoding, and secondly outlining what conditions and situations for which fuzzy logic techniques are most suited for these approaches.
3. A detailed account of some applications of fuzzy systems in communication is provided in this report. One of the earliest reported applications of fuzzy systems in wireless communication in [33]. It reports the use of an RLS fuzzy adaptive filter for non-linear channel equalization.
4. A work of similar nature that can also be considered to be among the few earliest reported is [34] and it deals in some detail with a fuzzy based channel equalization problem.
5. Another contemporary work is [35] which shows the use of fuzzy systems to carry out channel estimation. A similar survey paper is [36].
6. Another work [37] reports the use of fuzzy systems to perform channel estimation in CDMA based wireless communication.
7. A simple method reported in [38] shows that data sequence and estimates of the channel condition can be carried out at the same time using the Viterbi algorithm and fuzzy logic for the convolutional code. After a fixed number of decoding steps, the fuzzy logic unit reads the branch metric value of the survivor and the difference between maximum and minimum survivor path metric values at the Viterbi decoder and estimates the channel condition with the signal-to-noise ratio (SNR). The proposed method enables the channel estimation regardless of what kinds of modulator and demodulator are used.
8. Another work referred in [39] presents the equalization of channel distortion by using NF network. The structure and learning algorithm of NF network have been described. Using learning algorithm of NF network an adaptive equalizer have been developed. The developed equalizer recovers transmitted signal efficiently. The use of NF equalizer in digital signal transmission allows to decrease training time of parameters and the complexity of network.
9. A work cited in [40] discusses about a Takagi Sugeno Kang (TSK) fuzzy approach to channel estimation for OFDM systems.

10. Fuzzy systems can be used to update LMS algorithm for OFDM channel estimation in time-variant mobile channels. Such a work is reported in [41].
11. The workers of the publication cited in [40] extends the work further using a TSK fuzzy approach to channel estimation for MIMO-OFDM systems [42].
12. An adaptive NF inference for OFDM channel estimation is reported in [43].
13. Use of fuzzy logic as the core of the reasoning engine to determine different parameters used by the WiMAX system is reported in [44]. This work focuses on one of the main functions of the reasoning engine i.e. determination of the channel type and the number of pilots used for channel estimation.
14. Another work introduces an adaptive neural fuzzy channel equalizer (ANFCE) based on adaptive neural fuzzy filter (ANFF) [45]. The ANFF is a five layer ANN which is able to use the expert knowledge in its structure. The structure and parameters of this network are adjusted according to the training data and the available expert knowledge.
15. A work cited in [46] propose a computationally efficient NF system based equalizer for use in communication channels. This equalizer performs close to the optimum maximum a-posteriori probability (MAP) equalizer with a substantial reduction in computational complexity and can be trained with a supervised scalar clustering algorithm.
16 The work [47] proposes an adaptive NF inference system (ANFIS) for channel estimation in OFDM systems. To evaluate the performance of this estimator, the authors compare the ANFIS with LS algorithm, MMSE algorithm by using BER and mean square error (MSE) criteria.
17. The authors report the design of a fuzzy MLP for application in stochastic MIMO channels [48].

Here, we noticed that the major literature have been restricted to the popular NFS or FNS learning and decision-making arena with the focus to improve performance of such systems with applications in time-varying MIMO channel modeling and wireless communication. These works have highlighted how fuzzy based systems are able to deal with the uncertainty observed in wireless channels and also track minute variations which are created due to the time dependent nature of such channels.

16.6 Conclusion

Here, we have discussed about the application of soft computing tools for wireless communication applications. We focussed on the use of ANN in both feedforward and recurrent forms for dealing with a range of issues like channel equalization and estimation, interference cancelation, user identification etc. related to wireless communication. Fuzzy systems are able to deal with uncertainty, hence are useful for dealing with the stochastic nature observed in wireless channels. Fuzzy in combination of ANN form constitute a reliable framework for application in wireless channel.

Some recent works reported in this arena has been highlighted in this review. All the reported works, in totality, indicate that soft computing frameworks in form can be effective ingredients of receiver design suitable for data intensive mobile applications.

References

1. Haykin S (2006) Neural networks, 2nd edn. Pearson Education, New Delhi, India
2. Ling Z, Xianda Z (2007) MIMO channel estimation and equalization using three-layer neural networks with feedback. Tsinghua Sci Technol J 12(6):658–661
3. Muhammad J (2007) Artificial neural networks for location estimation and co-channel interference suppression in cellular networks. Master of Philosophy Thesis submitted to University of Stirling
4. Fang Y, Chow TWS (1999) Blind equalization of a noisy channel by linear neural network. IEEE Trans Neural Networks 10(4):918–924
5. Naveed A, Qureshi IM, Cheema TA, Jalil A (2004) Blind equalization and estimation of channel using artificial neural networks. In: Proceedings of INMIC. Lahore, Pakistan, pp 184–190
6. Zamiri-Jafarian H, Gulak G (2005) Iterative MIMO channel SVD estimation. In: Proceedings of IEEE international conference on communications. Seoul, Korea, pp 1–5
7. Zamiri-Jafarian H, Gulak G (2005) Adaptive channel SVD estimation for MIMO-OFDM systems. In: Proceedings of 61st IEEE semiannual vehicular technology conference. Stockholm, Sweden, pp 240–245
8. Ping W, Lhua L, Ping Z (2007) Subchannel interference cancelation in SVD-based MIMO system. J China Univ Posts Telecommun 14(3)
9. Ibnkahla M (2007) Neural network modeling and identification of nonlinear MIMO channels. In: Proceedings of 9th international symposium on signal processing and its applications (ISSPA), UAE, pp 1–4
10. Sarma KK, Mitra A (2009) ANN based Rayleigh multipath fading channel estimation of a MIMO-OFDM system. In: Proceedings of 1st IEEE Asian Himalayas international conference on internet (AH-ICI). Kathmandu, Nepal, pp 1–5
11. Hua C, Xiao-Hui Z (2010) MIMO-OFDM channel estimation based on neural network. In: Proceedings of WiCOM-2010. Chengdu, China, vol 6, pp 1–4
12. Wenle Z (2010) MADALINE neural network for parameter estimation of LTI MIMO systems. In: Proceedings of 29th Chinese control conference. Beijing, China, pp 1346–1351
13. Bhuyan M, Sarma KK (2012) MIMO-OFDM channel tracking using a dynamic ANN topology. World Acad Sci Eng Technol Online Int J 71:1321–1327
14. Sarma KK, Mitra A (2012) MIMO Channel modeling using FIR and IIR MLP topologies. In: Proceedings of international conference on intelligent infrastructure. Kolkata, India
15. Paul JR, Vladimirova T (2008) Blind equalization with recurrent neural networks using natural gradient. In: Proceedings of ISCCSP-2008. Malta, pp 178–183
16. Potter CG (2008) Multiple input multiple-output wireless communications with imperfect channel knowledge. Missouri University of Science and Technology, USA. http://proquest.umi.com
17. Potter CG, Venayagamoorthy GK, Kosbar K (2010) RNN based MIMO channel prediction. Elsevier J Sig Proc 90(2):440–450
18. Mao W (2008) Novel SREKF-based recurrent neural predictor for narrowband/FM interference rejection in GPS. Elsevier Int J Electr Commun 62:216–222
19. Paul JR, Vladimirova T (2008) Blind equalization with recurrent neural networks using natural gradient. In: Proceedings of ISCCSP-2008, Malta, pp 178–183
20. Li Y, Deng Y (2008) Nonlinear equalization with known channel state information in satellite communication. In: Proceedings of ICCS 2008. Guangzhou, China, pp 1081–1085

21. Coelho PHG (2002) Adaptive channel equalization using EKF-CRTRL neural networks. In: Proceedings of international joint conference on neural networks. Honolulu, Hawaii, pp 1195–1199
22. Gowrishankar Satyanarayana PS (2007) Recurrent neural network based BER prediction for NLOS channels. In: Proceedings of 4th international conference on mobile technology. Applications and systems, Singapore, pp 410–416
23. Choi J, Bouchard M, Yeap TH, Kwon O (2004) A derivative-free kalman filter for parameter estimation of recurrent neural networks and its applications to nonlinear channel equalization. In: Proceedings of 4th international ICSC symposium on engineering of intelligent systems (EIS 2004). Madeira, Portugal, pp 1–7
24. Choi J, Ko K, Hong I (2001) Equalization techniques using a simplified bilinear recursive polynomial perceptron with decision feedback. In: Proceedings of international joint conference on neural networks, Washington DC, USA, vol 4, pp 2883–2888
25. Choi J, Lima AC, Haykin S (2005) Kalman filter-trained recurrent neural equalizers for time-varying channels. IEEE Trans Commun 53:472–480
26. Choi J, Bouchard M, Yeap TH (2005) Decision feedback recurrent neural equalization with fast convergence rate. IEEE Trans Neural Netw 16:699–708
27. Liu W, Yang L, Hanzo L (2006) Recurrent neural network based narrowband channel prediction. In: Proceedings of 63rd IEEE VTC 2006. Melbourne, Australia, pp 2173–2177
28. Park D (2008) Equalization for a wireless ATM channel with a recurrent neural network pruned by genetic algorithm. In: Proceedings of 9th ACIS international conference on software engineering, artificial intelligence, networking, and parallel/distributed computing. Phuket, Thailand, pp 670–674
29. Gogoi P, Sarma KK (2012) Kalman filter and recurrent neural network based hybrid approach for space time block coded-MIMO set-up with Rayleigh fading. In: 13th WSEAS international conference on neural netwroks (NN 12), Isai, Romania
30. Mitra S, Hayashi Y (2000) Neuro-fuzzy rule generation: survey in soft computing framework. IEEE Trans Neural Netw II(3):748–768
31. Mitra S, Konwar KM, Pal SK (2002) Fuzzy decision tree, linguistic rules and fuzzy knowledge-based network: generation and evaluation. IEEE Trans Syst Man Cybern-Part C: Appl Rev 32(4):748–768
32. Erman M, Mohammed A, Rakus-Andersson E (2009) Fuzzy logic applications in wireless communications. In: Proceedings of IFSA-EUSFLAT. Lisbon, Portugal, pp 763–767
33. Wang LX, Mendel JM (1993) An RLS fuzzy adaptive filter with applications to nonlinear channel equalization. In: Proceedings of 2nd IEEE international conference on fuzzy systems. San Francisco, USA, vol 2, pp 895–900
34. Sarwal P, Srinath MD (1995) A fuzzy logic system for channel equalization. IEEE Trans Fuzzy Syst 3(2):246–249
35. Arnai F, Coulton P, Honary B (1995) Real time channel estimation based on fuzzy logic. In: Proceedings of IEEE international symposium on information theory. St.-Petersburg, Russia, pp 289–298
36. Ghosh S, Razouqi Q, Jerry Schumacher H, Celmins A (1998) A survey of recent advances in fuzzy logic in telecommunications networks and new challenges. IEEE Trans Fuzzy Syst 6(3):443–447
37. Niemi A, Joutsensalo J, Ristaniemi T (2000) Fuzzy channel estimation in multipath fading CDMA channel. In: Proceedings of 11th IEEE international symposium on personal, indoor and mobile radio communications, vol 2. London, UK, pp 1131–1135
38. Young-Bae B, Takama Y, Hirota K (2002) Combined channel estimation and data decoding based on fuzzy logic. IEEE Trans Instrum Measur 51(2):342–346
39. Abiyev RH, Al-shanableh T (2007) Neuro-fuzzy network for adaptive channel equalization. LNCS Adv Neural Networks-ISNN 4492:241–250
40. Zhang J, He ZM, Wang X, Huang Y (2006) A TSK fuzzy approach to channel estimation for OFDM systems. J Electr Sci Technol China 4(2)

41. Wen J, Chang C, Lee G, Huang C (2006) OFDM channel prediction using fuzzy update LMS algorithm in time-variant mobile channels. In: Proceedings of IEEE 6th vehicular technology conference. Montreal, Canada, pp 1–5
42. Zhang J, He Z, Wang X, Huang Y (2007) TSK fuzzy approach to channel estimation for MIMO-OFDM systems. IEEE Signal Process Lett 14(6)
43. Nuri Seyman M, Taspinar N (2008) Channel estimation based on adaptive neuro-fuzzy inference system in OFDM. IEICE Trans Commun E91-B(7):2426–2430
44. Shatila H, Khedr M, Reed JH (2009) Channel estimation for wiMax systems using fuzzy logic cognitive radio. In: Proceedings of IFIP international conference on wireless and optical communications networks (WOCN'09). Cairo, Egypt, pp 1–6
45. Gharibi F, Jamjah JR, Akhlaghian F, Azami BZ (2010) An improved adaptive neural fuzzy channel equalizer. In: Proceedings of 18th Iranian conference on electrical engineering, Isfahan, Iran, pp 326–330
46. Sahu PK, Patra SK, Panigrahi SP (2010) Non-linear channel equalization using computationally efficient neuro-fuzzy channel equalizer. In: Proceedings of 18th Iranian conference on electrical engineering, Isfahan, Iran, pp 326–330
47. Nuri M, Seyman, Taspinar N (2010) Channel estimation based on adaptive neuro-fuzzy inference system in OFDM. IEICE Trans Commun E91.B(7):2426–2430
48. Sarma KK, Mitra A (2013) Stochastic MIMO channel modeling using FMLP based inference engine. Int J Inf Commun Technol 5(2)

Author Index

K.K. Sarma et al. (eds.), *Recent Trends in Intelligent and Emerging Systems*,
Signals and Communication Technology, DOI 10.1007/978-81-322-2407-5